A CULTURAL HISTORY OF MATHEMATICS

VOLUME 4

A Cultural History of Mathematics
General Editors: David E. Rowe and Joseph W. Dauben

Volume 1
A Cultural History of Mathematics in Antiquity
Edited by Michael N. Fried

Volume 2
A Cultural History of Mathematics in the Medieval Age
Edited by Joseph W. Dauben, Clemency Montelle, and Kim Plofker

Volume 3
A Cultural History of Mathematics in the Early Modern Age
Edited by Jeanne Peiffer and Volker R. Remmert

Volume 4
A Cultural History of Mathematics in the Eighteenth Century
Edited by Maarten Bullynck

Volume 5
A Cultural History of Mathematics in the Nineteenth Century
Edited by Tom Archibald and David E. Rowe

Volume 6
A Cultural History of Mathematics in the Modern Age
Edited by Tom Archibald and David E. Rowe

A CULTURAL HISTORY OF MATHEMATICS

IN THE EIGHTEENTH CENTURY

VOLUME 4

Edited by Maarten Bullynck

BLOOMSBURY ACADEMIC
LONDON • NEW YORK • OXFORD • NEW DELHI • SYDNEY

BLOOMSBURY ACADEMIC
Bloomsbury Publishing Plc
50 Bedford Square, London, WC1B 3DP, UK
1385 Broadway, New York, NY 10018, USA
29 Earlsfort Terrace, Dublin 2, Ireland

BLOOMSBURY, BLOOMSBURY ACADEMIC and the Diana logo are trademarks of Bloomsbury Publishing Plc

First published in Great Britain 2024

Series design by Raven Design
Cover image: A philosopher giving a lecture on the orrery by Joseph Wright of Derby (1734–97). Derby Museum © Christophel Fine Art / Universal Images Group via Getty Images

A catalogue record for this book is available from the British Library.

A catalog record for this book is available from the Library of Congress.

ISBN: HB: 978-13500-6305-1
ePUB: 978-13502-9485-1
ePDF: 978-13502-9484-4
Set: 978-13500-6312-9

Series: The Cultural Histories Series

Typeset by RefineCatch Limited, Bungay, Suffolk
Printed and bound in Great Britain

To find out more about our authors and books visit www.bloomsbury.com and sign up for our newsletters.

CONTENTS

ILLUSTRATIONS

Introduction

Chapter 2

Chapter 3

Chapter 6

Chapter 7

CONTRIBUTORS

David Bellhouse is Professor Emeritus of Statistics at the University of Western Ontario, Canada. He holds a BA and MA from the University of Manitoba and a PhD from the University of Waterloo, Ontario, Canada. He has published several research papers on the history of probability and statistics as well as two books: *Abraham De Moivre: Setting the Stage for Classical Probability and Its Applications* (2011) and *Leases for Lives: The Emergence of Actuarial Science in Eighteenth-Century England* (2017). David is a former president of the Statistical Society of Canada and has served on the editorial boards of five statistical journals.

Franziska Bomski is a literary scholar whose work focuses on the intersection of literature and science, and the practice of interdisciplinarity in the humanities. Following a degree in German literature and mathematics, she earned a PhD from Freiburg University, Germany, with a study on mathematics in German romanticism. Franziska has also examined the literary significance of mathematics in the works of Robert Musil. Franziska taught German language and literature at Frankfurt University, Germany, Yale University, and Washington University, St. Louis, Missouri, USA, and Beihang University, Beijing, China. From 2012 to 2018, she served as the research coordinator at the Klassik Stiftung Weimar. In 2018, she joined the Einstein Forum in Potsdam, Germany. She is the author of *Die Mathematik im Denken und Dichten von Novalis* (Mathematics in the Mind and Poetics of Novalis, 2014) and co-editor of *Ordo inversus. Formen und Funktionen einer Denkfigur um 1800* (Ordo inversus: Forms and Functions of a Figure of Thought around 1800, 2020).

Maarten Bullynck studied mathematics, Germanic languages, and media theory at Gent University, Belgium, and Humboldt University of Berlin, Germany. His

PhD focused on the shifting structures of scientific discourse under the influence of changing communication patterns in the eighteenth century. Maarten is currently Maître de Conférences (Associate Professor) at the Department of Mathematics and History of Sciences of the University Paris 8, and member of IDHE.S (Institutions et Dynamiques Historiques de l'Économie et de la Société, UMR 8533), France. His research interests are the history of mathematics and computing in the eighteenth and twentieth centuries, focusing on computational and information processing practices and on how the everyday work of the mathematician is shaped and slowly changes through time. He recently published on the use of paper tools in mathematics, and on the early history of operating systems, publishing a chapter on "What is an Operating System? A Historical Investigation (1954–1964)" (2017). He co-edited with Liesbeth De Mol and Marie-José Durand-Richard a special issue on "Mathematics and Machines" in *Revue de Synthèse* (2018).

Jeanne Peiffer has made her career as a research scientist at the Centre National de la Recherche Scientifique (CNRS), and was affiliated with the Centre Alexandre Koyré, Paris, France, where she is still active as emerita. As a historian of mathematics, Jeanne focused on early modern mathematics, namely on Albrecht Dürer's contributions to geometry, publishing her book *Albrecht Dürer, Géométrie* (1995); and on Johann Bernoulli's correspondence, about which, see *Der Briefwechsel von Johann Bernoulli*, 2–4 (1988, 1992, and forthcoming). Jeanne is also the author of a popular *Histoire des mathématiques* (1982, translated into Russian, German, American English, and Spanish). More recently, she is part of an international network devoted to the study of the circulation of mathematics within and via journals, cf. *Interplay between Journals at Various Scales*, special issue of *Historia Mathematica* (2018). For more details, see: https://cak.ehess.fr/membres/jeanne-peiffer.

Helmut Pulte is Professor of Philosophy at the University of Bochum, Germany. He studied mathematics, physics, and philosophy and earned his PhD at Bochum with a thesis on the foundation of mechanics in the works of Euler, Maupertius, and Lagrange. After a research period as an Alexander von Humboldt Fellow at the University of Cambridge, UK, he received his *venia legendi* in 2002. His research focuses on the theory, philosophy, and history of science, especially of physics and mathematics, as well as on the changing notions of science since the Enlightenment. He is author of *Axiomatik und Empirie. Eine wissenschaftsgeschichtliche Untersuchung zur Mathematischen Naturphilosophie von Newton bis Neuman* (2005). His recent publications include *The Reception of Isaac Newton in Europe*, co-edited with Scott Mandelbrote (2018) as well as a series of studies on Jakob Friedrich Fries, Isaac Newton, Hermann von Helmholtz, and Johannes von Kries.

Jip van Besouw is a postdoctoral researcher at the Centre for Logic and Philosophy of Science at the Vrije Universiteit Brussel, Belgium, where he received his PhD in philosophy in 2017. He works on the history of seventeenth- and eighteenth-century natural philosophy. Jip has specific interests in the development of experimental and mathematical scientific practices at the time, as well as in the epistemological and methodological discourses surrounding those practices. He published, "The Wedge and the vis viva Controversy," in the *Archive for the History of Exact Sciences* (2016) and has written extensively on William Jacob 's Gravesande and the Dutch Newtonians in a number of journals.

Jane Wess holds a physics degree and spent two years teaching mathematics before a long career at the Science Museum, London, UK, where she was Curator of Mathematics from 1991 until 2012. More recently, she has completed a doctorate at Edinburgh University, studying the use of mathematical instruments in exploration. Her particular interests are in experimental philosophy and mathematics in the eighteenth century, having worked on the King George III Collection of instruments, King's College London. Jane has published on several aspects on the history of mathematics, including navigation, mathematical learning tools, and logic demonstrators. She has published numerous articles on eighteenth-century mathematics in the *British Journal for the History of Mathematics* and the *BSHM Bulletin*. She is also the co-author with Alan Q. Morton of *Public and Private Science: King George III Collection* (1993).

SERIES PREFACE

Culture, history, and mathematics are words that seldom appear together. Yet, mathematical activities have played a pervasive role in diverse cultures throughout human history, as described in the present six volumes. These illustrate how individuals and societies have engaged with various manifestations of mathematical culture from antiquity to the present. Based on recent and ongoing scholarship, they provide a global overview of a highly diverse range of developments, though certainly nothing like a comprehensive picture. Editors and authors had to make difficult choices about which topics and cultures to include, and in doing so, they have sought to strike a balance between characteristic features of a period and particular individuals and events that had a deep influence or longer lasting impact.

Each volume addresses seven general themes in as many chapters. The first three chapters—Everyday Numeracy, Practice and Profession, and Inventing Mathematics—highlight the roles of those who used, taught, or created mathematics as it evolved over time. These chapters explore mathematics per se, whereas the three that follow—Mathematics and Worldviews, Describing and Understanding the World, Mathematics and Technological Change—consider the larger impact of mathematics on other dimensions of human culture. The final chapter—Representing Mathematics—is devoted to a broader understanding of mathematical ideas, activities, or artifacts in different cultures throughout the world. The periods covered are:

Volume 1: A Cultural History of Mathematics in Antiquity
(3000 BCE–500 CE)
Volume 2: A Cultural History of Mathematics in the Medieval Age
(500–1400)

Volume 3: A Cultural History of Mathematics in the Early Modern Age (1400–1687)
Volume 4: A Cultural History of Mathematics in the Eighteenth Century (1687–1800)
Volume 5: A Cultural History of Mathematics in the Nineteenth Century (1800–1914)
Volume 6: A Cultural History of Mathematics in the Modern Age (1914–2020)

General Editors: David E. Rowe and Joseph W. Dauben

Introduction

MAARTEN BULLYNCK

The period from the latter half of the seventeenth century to the end of the eighteenth century is an important period of transition. This long eighteenth century harbors many changes on a number of levels: political, economic, cultural, and scientific. While building upon the gradual evolutions of the sixteenth and seventeenth centuries, it is in the second half of the eighteenth century that the balance slowly tips and more systemic changes started and sped up. A world economy that increasingly globalized; transports and infrastructure that gained stability and traction; a printing press and its products that accelerated and reached a growing audience: all these developments were under way, but gained momentum in the late eighteenth century.

This momentum also had an impact on the evolution of ideas and norms regarding religion, science and society. This is especially true in the European realm, where the eighteenth century is often called the Age of Enlightenment, where enlightenment refers in particular to the changing attitudes towards religion. Religious dogmatism was increasingly confronted with critical reflection and the role of religion in society was put into question. Instead of religion, according to many, reason should now prevail in society and the public sphere, and the sciences were vaunted as showing the ways in which this could be realized. These developments led the way to a stricter separation between religion and the state in Europe. Such a separation was gradually realized in different forms, be it by an "enlightened" ruler such as the emperor Joseph II in Hapsburg Austria or by the French Revolution (1789–99). These developments led to a redefinition of the state's role, its administration and its influence on science and education.

The very notion of Enlightenment was very much a European creation and phenomenon, though it interacted subtly but decisively with the global network of trade and colonization that European states were developing in the eighteenth century. During the late seventeenth century, the so-called "Quarrel between the Ancients and the Moderns" had questioned whether the ancient Greeks and Romans were still the models to be followed and whether or not they had been surpassed. The advocates of the modern would eventually prevail and the idea of scientific and cultural progress driving society would be much elaborated in the eighteenth century. The contact with other cultures prompted the printing of early ethnographic and many popular travel descriptions, making the European readership acutely aware of the world around them, and self-aware as well. On the one hand, this put their own moral codes and histories in perspective, but, on the other hand, also motivated the development of new historical and cultural narratives that could legitimate Europe's growing role and leadership globally. Historiographical schemes in which mankind progressed from primitive, tribal communities to ever more complex societies with increased policing and administration became fashionable in political rhetoric and in philosophy. As a result, not only were some earlier periods in European history dismissed as being "dark" (think of the Middle Ages), but also other cultures and societies were considered less enlightened and developed than European society.

The sciences and mathematics formed part of this self-serving definition of Enlightenment as they were seen as important wheels of progress in this narrative. Many such developments influenced mathematics and its place in society. The use of mathematics as an aid for planning and executing the proposed reforms gradually augmented its social status. Aside from the more elementary parts of mathematics that were useful in daily practice, the more advanced parts came to be more and more in demand. Indeed, mathematics during the long eighteenth century was not only in transition socially, but also mathematically. For example, the advantages of symbolic algebra became fully accepted, and the development of the calculus brought vast new fields to fruition. The application of these new techniques to numerous phenomena of the physical and the social world convincingly showed their value. Consequently, finding a place for algebra and the calculus within mathematics shook up the traditional vision of mathematics as a whole, restructuring its body of knowledge. This stirred up much discussion on the nature and the foundations of mathematics that would continue well into the nineteenth century. It would also lead to new, slowly standardizing ways of teaching mathematics, both elementary and advanced (see Chapters 1 and 2).

WHAT IS "MATHEMATICS" IN THE EIGHTEENTH CENTURY? CHANGING DEFINITIONS AND DOMAINS

A sense of the general change in what mathematics stood for in the long eighteenth century can be obtained by comparing definitions of mathematics from the early eighteenth century to those appearing later on. Christian Wolff in his *Mathematisches Lexicon* (1716) discerned between a pure mathematics (*mathesis pura*) that, as the science of quantities, could be studied as such (*mathesis theoretica sive speculativa*), and mixed mathematics, that is, mathematics applied to objects of nature that can be measured (*mathesis mixta sive impura*). Further, there was also the practical mathematics (*mathesis practica*) of surveyors, craftsmen, and mechanics that applied some notions of mathematics, though they did so blindly and from memory. According to Wolff, these craftsmen often wrongfully usurped the title of *Mathematicus*, thereby diminishing its social and scientific status.

In the *Encyclopédie* (1751–72) of Denis Diderot and Jean le Rond d'Alembert, by contrast, mathematics was also defined as the science of calculable or measurable quantities but was now subdivided only into *mathématiques pures* and *mathématiques mixtes* (i.e., as applied to concrete quantities). Practical mathematics did not figure. Under *géométrie*, practical geometry was briefly alluded to, but these practitioners did not need theory, certainly no rigorous demonstration, only some "*raisonnements*" (reasoning) to avoid them getting stuck in "*routine aveugle*" (blind routine). As the subdivision between pure and mixed would be recast in the nineteenth century as pure and applied mathematics, practical mathematics would disappear altogether. This shows clearly an increasing professional and scientific tension between practitioners of mathematics and, for lack of a better word, *professionals* of mathematics. The birth of mathematics as a discipline in the new educational system of the nineteenth century (see Volume 5) would close off this evolution, drawing a clear, institutional line between the professional mathematician and the others that merely used mathematics in their trade, profession or scientific discipline.

This shift in the professional status of mathematics not only created social tensions, but it also surfaced with epistemological overtones in the discourse on mathematics. In his *Lexicon*, Christian Wolff illustrated the difference between the professional and the occasional user of mathematics with the example of the astronomer Jean Picard's new leveling instrument. According to Wolff, the craftsmen lacked theory to understand and correct what they were doing with their instruments. This led them into making errors. During the eighteenth century, errors were normally understood as the result of a lack of knowledge. They were in that sense both a moral and a scientific category. Scientifically, an error might come from not knowing all circumstances; morally, an error might be born from a lack of understanding or from incomplete understanding. However, identifying and minimizing errors became increasingly important during the eighteenth century and the quantification of error, as part of probabilities, slowly made it possible to estimate its impact (see Chapter 4). In Wolff's example, the hidden errors of leveling instruments could be corrected if mathematical theory was added to practice, supplying the causes unknown

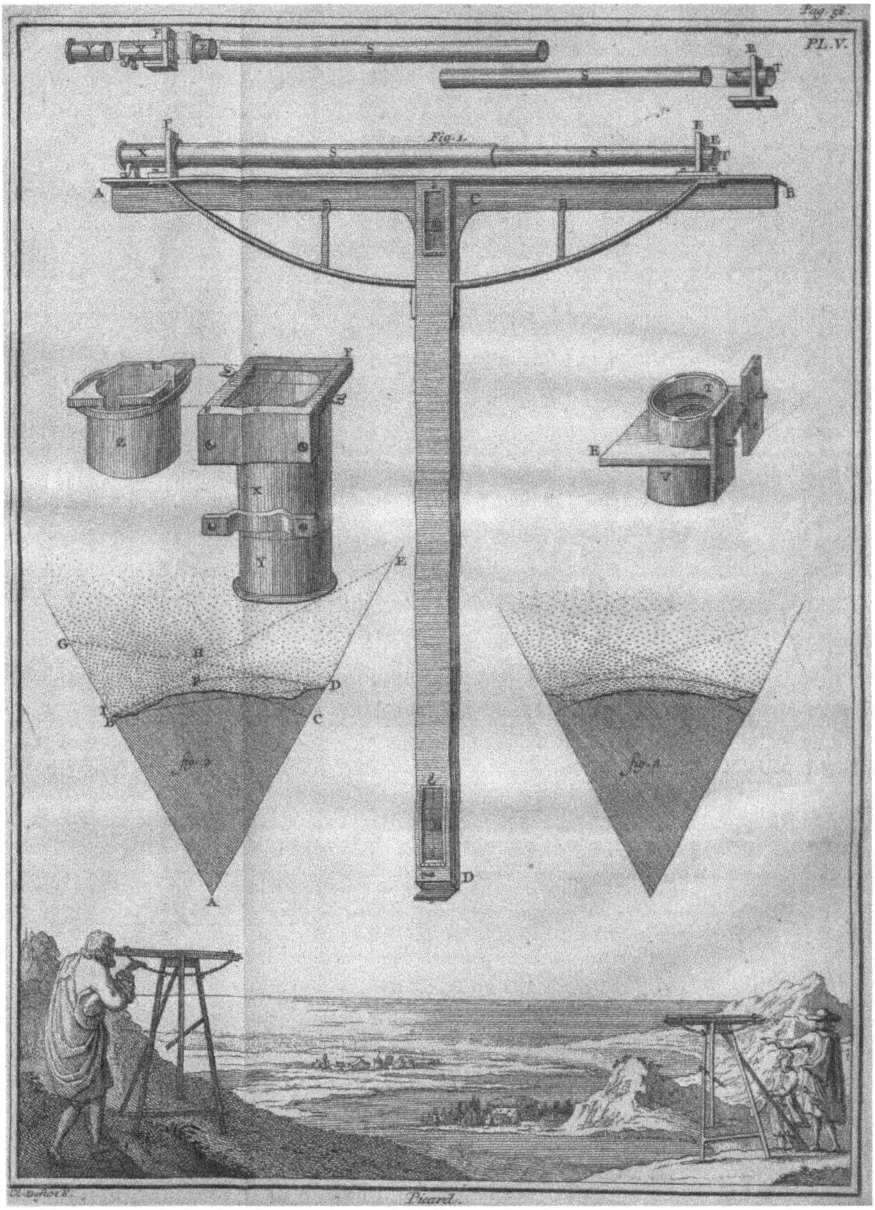

FIGURE 0.1. Picard's levelling instrument (1671). Photo courtesy of ETH-Bibliothek Zürich, Rar 4358, Public Domain.

beforehand that led to error. The ignorance of the craftsmen using only practical mathematics led them into error and set them apart from the professional of pure and mixed mathematics: the true Mathematician who could steer his knowledge to find the errors and set them straight.

The details of Wolff's exemplary case, Picard's theory of the level, are worth elaborating, as they paradigmatically bring out the dimensions of this divide. Picard had shown in a treatise that was published in 1684 by Philippe de la Hire that for distances of over 100 *toises* (approximately 2,000 meters), the apparent leveled distance arising from the use of the instrument was in error with respect to the real leveled distance, an error increasing with larger distances. This error was due the curvature of the earth, and for a distance of 100 *toises* the error was of about 5 *lignes* (approximately 1 centimeter). This problem hints at two aspects of professionalization. First, it is symptomatic of the scaling up of mathematical practices, from local surveying that can ignore such errors to inter-regional and coordinated endeavors of drawing maps, determining national borders and so forth (see also Chapter 2). Second, the error can only be derived when combining a local measurement with the global theory of the shape of the earth, a feat made possible by the success of rational mechanics (see Chapter 5). To determine (or approximate) the error quantitatively required more advanced mathematics, in this case trigonometry. Both the know-how and the knowledge of the practitioner were thus challenged by the upwardly mobile professional. The transition was, of course, often much smoother "in the field," outside the philosophical discourse of Wolff. To accommodate the practitioner in the field, Picard had included a table to make the calculation of the error easy, even without knowledge of trigonometry by simple lookup. He also added a method of leveling (viz., measuring once from A to B, then from B to A) that automatically adjusted the distance.

This renegotiation of social and professional status in association with a shifting systematics of mathematical knowledge formed an integral part of the transitional character of the eighteenth century in the history of mathematics.

THE EIGHTEENTH CENTURY OF BATTLEFIELDS AND MARKETPLACES

The end of the Thirty Years' War in 1648, the end in England of Cromwell's rule in 1658, and the Glorious Revolution of 1688 mark the end points of a string of European wars that were religiously motivated and had upturned both the political order and the general population. The Thirty Years' War (1618–48) was the most impactful of these, ending in massacres, devastation of towns and villages, and plague, killing nearly a third of the population in many parts of Central Europe. The long eighteenth century would have numerous wars also, but they would generally be devoid of religious motives and were instead the symptom of ongoing power struggles between established and upcoming nations wanting to extend or defend their areas of influence. These wars aimed at controlling important passageways of trade like the North Sea or the Baltic Sea, at extending the realm of empire within Europe, or were battles over the globally distributed assets and rights of the colonial powers Spain, the Netherlands, France, and the United Kingdom.

During these conflicts, many states slowly developed a fiscal-military apparatus to support continuing struggles for power (Brewer 1989). Professional armies were trained and maintained, financed on an ongoing basis through the modernization and systematization of taxation. The local populations would no longer be the direct, physical victims of the hostilities (as had often been the case in the seventeenth century), though they suffered indirectly because more efficient bureaucracies now implemented new taxation systems to pay for the wars. Similarly, the direct approach to colonization through settlement and plantations (as had been typical for the seventeenth century Caribbean islands) would also be supplanted slowly by a system where the military and the bureaucracy were deployed to control the economies abroad, a system perfected by the British Empire in the nineteenth century.

As John Brewer remarked in his study of eighteenth-century Great Britain, “power followed profit into the eighteenth century.” Indeed, exploitation of natural resources, control of trade routes or organization of taxes were increasingly important in shaping the activities of nations and of companies. The management of the eighteenth-century state was deeply intertwined with the actions of trading companies, with manufacturing and with the exploitation of land and other natural resources such as forests and mines. Many of the bigger enterprises of the time were forms of public–private partnerships, often through the granting of concessions, privileges, and monopoly. The Vereenigde Nederlandsche Oost Indische Compagnie was a Dutch state-owned company, the East India Company obtained exclusive privileges to trade with India by repeatedly loaning the British Crown millions of pounds, and the shares of France’s Compagnie perpétuelle des Indes were used (unsuccessfully) by John Law to redress France’s debts. These intricate financial constructions finally led to the first financial crises, notably the Mississippi Bubble in Paris, followed by the South Sea Bubble in London (1720).

In many respects, the Seven Years’ War (1756–63) was the epitome of the eighteenth-century war. Although the focus of the war seemingly was Central Europe, with Frederick the Great trying to expand Prussia, the war was concurrently fought around the Baltic Sea, North America, and the Indies. It may been considered one of the first “world wars.” The war ended with France losing an important part of its colonies to England. This started an overhaul of France’s finances. The French renegotiated their debts by paying flat-rate annuities to the Geneva banks, and they began reorganizing the tax system and the bureaucracy

FIGURE 0.2. The Bubblers Medley, or a Sketch of the Times', 1720. Satire on the South Sea Bubble of 1720. Photo by Guildhall Library & Art Gallery / Heritage Images / Getty Images.

that was to implement it. The defeat also prompted a reform of the military that met with much resistance from the privileged classes (see Chapter 2). Though Prussia successfully established itself as a major force in Central Europe, it also left the war with considerable debts that led to a financial crisis. This crisis started in Amsterdam, then the heart of worldwide banking. Some banks had speculated on the loans to Prussia and now failed, with the crisis spreading to Hamburg and other financial centers. As in the war itself, London was the winner, becoming the financial center of the world for the next two centuries, replacing Amsterdam. The Seven Years' War had also broken diplomatic new ground, shifting alliances that had been steady over decades, and had reshuffled the power balance, not only in Europe, but also globally, between colonial powers and within the colonies. In the aftermath of the Seven Years' War, as Great Britain sought to recover from the expenditures of the war by imposing new and higher taxes on their colonies, the American Colonies declared independence, leading to the Revolutionary War and the foundation of the United States of America.

The most direct influence of these wars on mathematics was in military education. Successes or failures of past wars inevitably led to reorganizing the training of officers. When it became clear that since Louis XIV the French could rely on excellent topographic maps made by their *ingénieurs-géographes* (engineers-geographers) and that these maps were an integral part of their military success, the surveying curriculum of officers was duly updated in the other countries. After the naval defeats during the Seven Years' War, France reformed its military schools to catch up with the UK. Its example was followed throughout Continental Europe where new military schools were established in the late eighteenth century (see Chapter 2). Although gunnery largely remained in the realm of the empirical and the practical, mathematics-based ballistics, informed by field experiments, came to be used near the end of the eighteenth century (see Chapter 6).

More indirectly, the economic and financial stakes at play, as well as the financial constructions needed to uphold the war effort, also made use of the services of mathematics. The advanced financial technology prompted a need for more mathematical refinement, most obviously in the calculation of annuities (see Chapter 4), but also for grounding a more "rational" economics on a mathematical basis as for instance the French Physiocrats advocated. On a much broader scale, the organization of a fiscal-military apparatus mobilized traditional parts of mathematics that now had to be scaled up. The reform of land taxation called for better and more coordinated surveying. The traditional local surveying practices had to be integrated into the grander project of creating a nationwide overview of land and income (see Chapter 2). Managing taxes rationally also meant securing a stable income for the state through taxation with checks and balances. Taxes on sugar and alcohol were an important revenue source, and the excise men were one of the largest groups conversant in mathematics throughout the century (see Chapter 6).

As was already clear from its role in underwriting the financial constructions of the day, intercontinental trade was big business. Goods were shipped mainly between Europe and the Americas, but also increasingly to Southeast Asia, India, and, near the end of the eighteenth century, also to Australia. The west coast of Africa was also bound up in the trade routes, as part of the slave trade routes. The slave trade more than tripled in the eighteenth century. In the seventeenth century, an estimated 1.5 million people were deported from their homeland to become slaves in the European colonies, between 1700 and 1800 the slave trade grew to an estimated 5 to 6 million people. The main destination were the slave-based plantation economies in the Caribbean that were Europe's most lucrative colonies during this period, exporting sugar and coffee to the "motherland."

It was thus pivotal that the ships made their trips as safely and speedily as possible so as to increase and maximize profits. Here, the mathematical sciences could contribute by improving ship design and methods of navigation (see Chapter 6). The problem of finding longitude at sea, for which a British Board of Longitude had set a price in 1714, attracted particular interest. Mathematicians such as Leonhard Euler or Tobias Mayer, conducted detailed work on lunar theory to compute tables of the position of the moon that could be used in the lunar distance method of computing longitude.

Intercontinental trade had also made the world bigger. Not only were goods and people shipped from Europe to Africa, the Americas and Asia, but also technological, practical, and theoretical knowledge. Moreover, European countries collected specimens from all over the world: animals, plants, and artifacts were scouted and put in private and public collections. In this context, Steven J. Harris (1999) speaks of Europe in the eighteenth century as being a net exporter of manufactured goods, including texts, instruments, and the like, and a net importer of "raw" materials, such as natural objects and observations.

This also translated to another level, that of long and sometimes exotic voyages that were recounted in detail, illustrated with engravings in books and journals, opening up the world for the eighteenth-century readers. Travel literature and their fictionalizations in novels (for example, *Robinson Crusoe*) appeared, satisfying the appetite of the European reader for knowing the world outside. Scientific travels equally aroused much curiosity, be it travels to gather new specimens of plants or animals, or travels to map and measure the world. Pierre Louis de Maupertuis and Alexis Clairaut's expedition to Lapland (1736–7) and Charles Marie de La Condamine and Pierre Bouguer's expedition to Peru (1735–45) are two major instances of such great expeditions, organized by the Paris Academy, to measure large meridian arcs to determine the figure of the earth. These expeditions resonated in many contexts: the scientific results were published in the proceedings of the Academy, and longer, more detailed accounts were published in book form afterwards. The proof, credited to

Maupertuis, that the earth was flattened at the poles, thereby proving Isaac Newton right and René Descartes wrong, found a wide public.

As for the export and import of mathematical knowledge, the Jesuits in particular had been exporting mathematics and its instruments (especially astronomical) to parts of India and to China since the seventeenth century. The big trading companies, too, had been exporting mathematical knowledge to the Americas and to Japan. The local uptake of this mathematical knowledge was quite limited. On the one hand, this cultural transmission was only an affair of the élites. On the other hand, while specific techniques and instruments were appropriated in their new locations, the frameworks in which they were embedded did not reproduce so easily. This is true in particular of the deductive structure of European mathematics and of the worldviews underlying the use of mathematics, be they heliocentric and Keplerian, Cartesian, or Newtonian. The place of mathematics in local worldviews and knowledge systems turned out to be quite incompatible with the European view of its importance. Though cultures with a rich mathematical tradition of their own, including the Muslim Ottoman Empire, India, and China, did publish, translate, and study some European mathematical texts in the 1700s, at the same time their practices remained grounded in their own tradition, even studying their old classics anew (Harris 1999).

There was also a reverse importation, as knowledge about Chinese or Indian culture and science was transmitted to Europe, mainly through the Jesuits. Accounts of the indigenous cultures of the Americas and of the Australasian archipelagoes made their way into Europe. This made a mark on the enlightened philosophers, opening up some kind of cultural relativism. Most famously, Gottfried Wilhelm Leibniz thought that the binary number system he had developed might have already been used for centuries by the Chinese. A deeper interest in scientific and mathematical texts from outside Europe, however, only developed late in the eighteenth century, when philologists and mathematicians would begin to study and compare Greek, Arabic, Indian, and Chinese sources more systematically.

THE REPUBLIC OF LETTERS AND THE OCEAN OF KNOWLEDGE

The eighteenth century witnessed a communication revolution. It built on seventeenth-century innovations such as the improved postal system and early periodical publications, but the exponential growth and spread only happened during the eighteenth century. The acceleration in spreading the written word and the growing public of readers (and listeners as texts were read out in public) made a deep imprint on eighteenth-century mentality. In this context, one often speaks of the "Republic of Letters," a virtual community transcending national borders and sharing a common culture of reading and writing. While this

"Republic" had some reality for the élite of the readership, for most people living in the 1700s this manifested itself largely as a rich exchange of news in multiple forms. Such news informed and misinformed their everyday outlook on life and the world. Indeed, the appearance and circulation of wildly diverse facts and fictions, gazettes and novels took up pace in the eighteenth century. On street corners, during meetings in the parks, or at tables in coffee or tobacco houses, news was read out aloud, transmitted orally and heatedly discussed. The dissemination of mail and the press opened up the world but also increasingly drowned the casual reader in an "ocean of knowledge." It is this distinction that is voiced by John Locke in his "Conduct of the Understanding":

> They have a pretty traffic with known correspondents, in some little creek; within that they confine themselves, and are dexterous managers enough of the wares and products of that corner, with which they content themselves, but will not venture out into the great ocean of knowledge. (Locke 1706a: 10)

The postal system had been expanding since the seventeenth century, when states (UK, France) or big companies (Thurn and Taxis), by monopoly, had taken the organization of the mail in hand. The number of postal stations grew exponentially during the eighteenth century, streamlining not only the routes and timetables of the post, but also structuring traffic by building new and better roads and turnpikes between post stations (see also Chapter 6). This postal road system was not only the vehicle for sending letters, but also for persons who were traveling in the postal stagecoaches, thereby opening up Europe for travel and fast communication. The impact of this road network continued to be felt well into the twentieth century: railroads followed postal roads, and telegraph and telephone lines followed railroads. Thus, the eighteenth-century communication network became entrenched as the backbone of modernity.

This quite reliable postal system was the material basis for the birth of periodicals that would start to appear in the seventeenth century. The first periodical publications were *gazettes* or *Zeitungen* whose appearance followed the rhythm of the postal service. They collected and edited news from all over, using the postal system, and quickly become popular. People like Henry Oldenburg, Otto Mencke, Denis de Sallo, Pierre Bayle, and many other information brokers were connected in a network of letters with many learned men across Europe. From these extensive personal correspondences evolved the first periodical publications of a more specific nature, such as scientific periodicals (see Chapter 2 and Peiffer and Vittu 2013), which started to appear shortly before 1700. In the course of the century, these correspondences developed into active networks in which information about the learned world was solicited and in which exchange of opinions and sometimes heated debate was conducted.

This exchange of letters between the learned was the lifeline of scientific communication in the eighteenth century. The very extensive correspondences among the learned testify to the ongoing communication in the Republic of Letters, trying to keep up with the latest developments. Each letter was, as it were, a node in an intricate web of writings, journals, books, other letters and travels. Typical letters reflected much on the practical circumstances that made the Republic of Letters possible. They often started by asking whether a person had seen or read this article or that book, inquiring whether the person could get hold of a book and then send it to the writer. They further talked of the latest publications, of what happened behind the scenes of journals or academies, excerpting parts of what the person had read, transmitting queries from other persons of the correspondence network, etc. The part of a letter that did not discuss the conditioning of accessing and transmitting knowledge, the "original" and personal part, was in itself also a transitory form. It was often a kind of "try-out" of ideas or of reworking readings that were still in the process of development before getting fixed in a more stable form. Eventually, such letters, often excerpted, would regularly be published in a journal, or be integrated into a larger text or book when the topic had reached a more definite form.

The commerce of letters also changed the form in which scientists discussed and argued with their colleagues in the public sphere (see Peiffer and Vittu 2013). Expressing a challenge in the public sphere, usual already in the sixteenth and seventeenth centuries, now became much easier. The famous mathematical challenges of the 1600s (Descartes, Fermat, John Pell, John Wallis, and many others) now became commonplace in the periodicals and the pace of rivalry now followed the frequency of the postal service and of monthly periodicals.

Especially in the development of the calculus, this format of scientific competition was often used and bore many fruits, both new results and bitter controversies. The publication of Leibniz's "Nova Methodus pro Maximis et Minimis" (1684) in the *Acta Eruditorum* is by itself an example of how the new communication form reformatted challenges (Leibniz 1684). The publication had been prompted by the fact that Leibniz's correspondent Ehrenfried Walther von Tschirnhaus might unveil part of Leibniz's discoveries in his own publications in the *Acta Eruditorum*. To foreclose the possibility of losing his priority of publication, Leibniz rather hurriedly drew up his article. Leibniz's text was brief and deemed enigmatic by most contemporaries, not least because of the many misprints. Thus it was in itself a challenge to their comprehension, just as Descartes' *Géométrie* had been a century and a half earlier. The paper further illustrated the method of differentials with four examples. Three stemmed from the seventeenth-century challenges that had circulated in the letters exchanged between Descartes, Pierre de Fermat, and Florimond de Beaune. By reposing these "classic" challenges in a new mathematical notation, the article proved that the new method could accelerate the solution of mathematical problems with a pedigree. It also transposed problems that had been discussed in personal correspondence to the new format of the journal, in the public sphere, where every learned person could read and scrutinize it. In the years afterwards, the pages of the *Acta* would regularly be filled with new problems and challenges set for others to be solved. Among the main mathematical actors were the brothers Jacob and Johann (Jean) Bernoulli, who developed Leibniz's technique and applied it to a variety of problems (see Chapter 3). Their own rivalry and their extensive correspondences with other mathematically literate persons throughout Europe were the source from which many of the public challenges arose.

This publishing activity on the side of the Bernoullis and their correspondents did not fail to arouse attention from across the English Channel. There, Newton's manuscripts on his technique of fluxions had circulated quite widely, but only within a select group, as had been usual in the seventeenth century. John Collins of the Royal Society and later William Jones, possessed a number of Newton's manuscripts and communicated their contents to people asking for them, the manuscript thus obtaining some semi-public status (Guicciardini 2009). As John Keill accused Leibniz of plagiarizing Newton in the *Philosophical Transactions* of 1708 and the infamous flurry of letters in the *Commercium Epistolicum* (1712) was published, the notorious priority quarrel over the invention of the calculus began its course (see Chapter 3). The debate was fought not only on the mathematical level, but also on the philosophical and even theological level, attracting much interest in the new calculus among members of the Republic of Letters (see Chapter 4). Periodical publications such as the *Acta Eruditorum*, the *Philosophical Transactions*, or the *Journal des Savants* were a preferred battleground for the debate, but it also found echoes in more general periodicals and in the yearly publications of the academies.

Indeed, the academies of science were drawn into this battle, too. While the Academies normally had a regulatory or even normative role in the Republic of Letters (see Chapter 2), discussions within the academies and among the academies were quite common, but most did not get out into the public sphere. The quarrel over the calculus, however, did engage both academies and the public sphere. The Royal Society sided with Newton after Leibniz, as a foreign member of the Society, had asked them to settle the matter. Other academies witnessed infighting between partisan parties, for or against Newton or Newtonianism. The Paris Academy experienced quarrels between Cartesians and Newtonians around 1700, where Guillaume de l'Hôpital, Pierre Varignon, and Joseph Saurin successfully defended the results of both Newton and Leibniz against criticisms from Michel Rolle. Maupertuis and Voltaire would later forcefully spread their interpretation of Newtonianism in France. In the middle of the eighteenth century, the Berlin Academy, originally formed according to a plan by Leibniz, also experienced a controversial episode dividing the academy. As the Newtonian Maupertuis was indirectly criticized for having derived his principle of least action from Leibniz, a long discussion ensued between the powerful Maupertuis, who was president of the Academy, and other members who saw themselves standing in the tradition of Leibniz and Wolff. Voltaire aptly satirized the whole episode in his *Histoire du docteur Akakia* (1752).

Claims of plagiarism or priority issues were fervidly discussed in the periodical literature. One must not forget that, though news traveled faster than before, there was no central authority that kept track of all publications internationally. Moreover, the ideas of originality or copyright either did not exist or did not have the current meaning. Many scientific publications of the eighteenth century, when seen through modern eyes, look like examples, commentaries, or emulations inspired by already published writings. Translating a geometric problem into algebra or bringing an annotated literature overview on a mathematical problem all constituted genuine and acceptable contributions to the scientific literature and possible stepping stones to a position at an institute for higher education or at a local academy. As a consequence, the literary and review journals very frequently had to address to issue of originality, sometimes starting off long and acrimonious discussions. As Albrecht von Haller described when he launched the review journal *Göttingische Gelehrte Anzeigen* in 1748, reviewers had to praise the good books, but publicly denounce

> those who attack God; thieves of the work of others; those who build castles in the sky, that possess the rule-based essence of mathematics without any solidity; those who repeat old and widely known truths, but present them as if they were their own and hide their sources.

In this realm of the "counterfeits of truth," those who abused mathematics figured explicitly. Not only could the truth and knowledge be a mere derivative of things already published elsewhere, but it could also, by design and construction, be falsehood and fable. Again, we see attempts to draw the line neatly between those who handled mathematics skillfully and professionally, and those who only had shallow notions of mathematicians and dabbled in speculation and metaphysics.

In this negotiation between different realms of users of mathematics, there were certain problems, even some with long pedigree, that were expelled from the professional domain and left to the amateurs. The bigger academies were the main institutions to enforce this discipline and even censorship on mathematical discourse. The most explicit case was that of the problem of squaring the circle, one of the classical Greek problems that had generated countless attempts at solution. In May 1775, the Royal Academy of Paris officially announced it would no longer examine any paper pretending to solve the quadrature of the circle (Jacob 2005). The problem of the duplication of the cube and the invention of a *perpetuum mobile* were equally banned. The

FIGURE 0.3. Frontispiece of Voltaire's *Elémens de la philosophie de Newton* (1738) showing how Newton from the heavens enlightens Voltaire. Photo by Culture Club / Getty Images.

reason was not so much mathematical, as most mathematicians at the Academy had been convinced of its impossibility since the 1700s, but rather social. There had been a growing number of amateurs of mathematics that had sent papers on these topics to the Paris Academy. Between 1740 and 1770, over a hundred attempts were submitted, the surge connected to a rumor that a prize had been created for solving the problem. Despite the Academy's decision not to waste any more time on examining proposed solutions, the amateurs persisted, but now published in local periodicals or self-published their tracts instead. A similar fate befell the attempts to find general method to solve equations of the fifth or higher degree at the Berlin Academy after Joseph-Louis Lagrange published his 1770–1 memoir that strongly suggested this was impossible.

FROM ENCYCLOPEDIAS TO SYSTEMS: WRITING MATHEMATICAL BOOKS

Book production tripled between the early 1700s and the end of the eighteenth century. For each of the national centers of publishing, be it London for the English market, Paris for the French market or Frankfurt for the German market, one sees the number of publications going up from some 1,000 titles per year in 1700 to over 3,000 per year in 1800. The main increase happened in the latter third of the century, energized by new technological developments in printing and publishing, and fed by an exponentially growing readership. The number of publications was such that a market for review journals emerged. These summarized and criticized the latest works, serving as a compass for the interested reader. In the last third of the eighteenth century, many reading clubs were founded, where by paying a small fee, one could peruse the books and journals bought by the club for its library. Next to the daily paper, the review journals belonged to the essential inventory of such clubs as they provided the latest update on the world of the book. It did not take long before the people who read only by proxy, that is, reading secondhand accounts and reviews instead of the books themselves, were duly satirized by philosophers and writers who complained of the "review factories" (after Adolph Freiherr von Knigge).

Indeed, access to books, either in private or "institutional" libraries (e.g., royal libraries or university libraries), in reading clubs or through proximity to the publishing houses (Paris, London, Amsterdam, but also Frankfurt and Leipzig), became more and more a necessity for keeping up in the Republic of Letters. The learned practice of collecting excerpts from books and ordering them according to topics, the so-called commonplace method of the seventeenth century, acquired a new dimension through the increasing volume of publications that could be accessed (Décultot 2003; Blair 2021). This mass of works provided the material basis of the encyclopedic format, multi-volume books that contained near exhaustive knowledge of a specific field in an orderly, mostly alphabetical, manner. From Pierre Bayle's *Dictionnaire historique et critique* (4 vols, 1697) over John Harris' *Lexicon Technicum* (2 vols, 1704–10) and Ephraim Chambers' *Cyclopedia* (2 vols, 1728) or Johann Heinrich Zedler's *Grosses vollständiges Universal-Lexicon aller Wissenschafften und Künste* (68 vols, 1731–54) to d'Alembert and Diderot's *Encyclopédie ou Dictionnaire raisonné des sciences, des arts et des métiers* (28 vols, 1751–72), and the *Encyclopédie Méthodique* (206 vols, 1782–1832), the eighteenth century was rife with encyclopedic ventures (see also Chapter 7). The majority of their contents was hardly original. Entries were copied and excerpted mostly verbatim from a number of sources, often enriched with commentaries and supplemented with more quotes. This did not diminish their value, on the contrary, it allowed a unique access to a rich repository of knowledge, framed and fitted in a system, larded with well-informed opinions and well-articulated viewpoints. Collating and compressing all (bookish) knowledge within a field helped to both stabilize a field of inquiry and delineate it from others. For example, one volume of the *Encyclopédie méthodique* (1799) even ventured to bring the theory of probabilities and all mathematical knowledge about games together.

The same objectives drove the publication of many specialized dictionaries, lexicons or manuals. For instance, dictionaries of mathematics were published by Jacques Ozanam (1691), John Harris (1704), and Christian Wolff (1716). These three authors were no novices when it came to collecting and systematizing mathematical knowledge. They were all "literary entrepreneurs," publishing widely on things both mathematical and not. All three, for instance, published a book on algebra. Harris' *Short Treatise on Algebra* appeared 1702, Ozanam's *Nouveaux élémens d'algèbre* (2 vols) also in 1702, Wolff's *Matheseos Elementa Universae* (2 vols) in 1713 to 1715.

Already near the end of the seventeenth century, the need for a well-organized and comprehensive treatise on algebra had been felt. The books published by John Pell (1668), John Kersey (1673–4), and John Wallis (1685) had tried to achieve this goal as early as the end of the seventeenth century (Pycior 1997; Stedall 2002). In France, the friars of the Oratorian order who were, like the Jesuits, much invested in education, provided for the publication of an algebra textbook, namely Jean Prestet's influential *Elémens des mathématiques* (1670, 2nd edition 1689). Indeed, there was a tendency among a growing number of authors to put algebra before arithmetic and geometry, as the more general way of doing things, or as the toolbox of discovery and invention for solving problems.

But it was in the eighteenth century that a variety of textbooks or manuals on algebra appeared, most of them in a vernacular language. Nearly all these manuals were written by people involved in teaching mathematics, though the books differed considerably in form and presentation according to the intended

FIGURE 0.4. Portrait of encyclopedist and mathematician Jean le Rond d'Alembert (1717–83) by Henriquez (1750). Surrounded by his books and equipped by the tools of the18th-century mathematician. On the table geometric instruments (compass, ruler, goniometer) and a number of loose notes, based on their size, letters d'Alembert is writing. Photo by SSPL / Getty Images.

audience and the local tradition. Going back to the examples cited earlier, Harris' *Algebra* was intended for his audience of his public mathematical lectures at Marine Coffee House in Birchin Lane, London, a course paid for by his patron George Cox. The book was focused on how to solve problems using algebra (the "algorithmic" side of algebra as one might say), but also included a short exposition on fluxions. Wolff's volume was meant for his students at the university and for the general learned public, and as such it stood more in the academic tradition. Inspired by Prestet's demonstrative presentation of algebra, Wolff streamlined it further, Euclidean style, and then showed how algebra could be applied to geometry and to both determinate and indeterminate problems in arithmetic. This rather encyclopedic volume continued with an exposition of trigonometric functions, and even of the principles of the infinitesimal calculus. Ozanam then was a private teacher and prolific author of mathematics. His two-volume treatise on algebra was in a way the counterpart to his popular *Récréations mathématiques et physiques* (1694). In his algebra, Ozanam was both systematic and problem-oriented, using algebra to rework hundreds of classic problems stemming from Diophantus and from the recreational tradition in mathematics. In particular, he devoted much attention to so-called indeterminate problems, trying to systematize heuristics and methods for solving problems of this kind. These three books may stand as a representative sample of algebra books for the eighteenth century, some aimed at practical use of (algorithmic) algebra, some at a systematized or encyclopedic presentation suited for universities, still others trying to show algebra's solving power for attacking problems.

The class of indeterminate problems have a special status here. These are problems where a solution had to be either a fraction or an integer and which could have either no solution, just one, or multiple solutions. Many stemmed from Diophantus' *Arithmetica*. Though algebraization of these problems often helped to solve them, there were no general or systematic methods of solution known, which made this class of problems challenging to explore. Indeed, they had already figured in Frans van Schooten's *Exercitationes mathematicae* (1657) a work that had introduced a generation of mathematicians (among them Newton and Euler) to fields where Descartes' analytical methods might be fruitfully applied. John Pell had tried his hand at it in 1668, emphasizing the importance of prime numbers and factorization for solving these problems either systematically or by trial-and-error. Later, algebra authors such as Wallis, Ozanam or Nicholas Saunderson pursued their integration into algebra, with mixed success. A breakthrough came when Euler, stimulated by his correspondence with Christian Goldbach, managed to solve larger classes of indeterminate problems. Euler, unfamiliar with an earlier result from Bachet, showed how the algorithm for finding the least common divisor of two numbers was pivotal for solving so-called Diophantine equations of the first degree. An example of such a problem asks us to find x and y such that $ax+by=c$ with x, y, a, b, c all integers. He further introduced continued fractions to solve Diophantine equations of the second degree (asking for example to find x and y such that $ax^2+by^2=c$ with x, y, a, b, c all integers). The growing body of results made Johann Heinrich Lambert lament in 1770 that he had found "isolated pieces, which did not seem possible to make easily into a connected and well-formed system" in the theory of prime numbers. Lambert subsequently launched a call to compute useful tables of factorization and to supply the missing theory. Indirectly, Euler responded to Lambert's activity and took up his old studies again in the 1770s. When d'Alembert asked Lagrange for a French translation of Euler's *Vollständige Anleitung zur Algebra* (1770), Lagrange suggested that the translation could be used as an opportunity to supply the "book on the science" of Diophantus' analysis where it is treated "methodically, explaining all known rules for the solution of indeterminate problems." The result became the second volume of Euler's *Elémens d'algèbre* (1774) with additions by Lagrange. This stands as the first textbook on what would later be called number theory. Adrien-Marie Legendre's *Essai d'une théorie des nombres* (1798) and Carl Friedrich Gauss' *Disquisitiones arithmeticae* (1801) would take off from where Euler and Lagrange left the emerging field in 1774 (Bullynck 2008b).

As mentioned before, first notions of fluxions or of differential calculus sometimes made their appearance under the heading of algebra, though mostly at the end of the textbooks, as a teaser for those wanting to study the newest topics further. If one wished to study infinitesimal calculus in more detail, one needed to find a private tutor, or one had to embark on reading articles and treatises as an autodidact (see Chapters 1, 2, and 3). The first manuals on the calculus appeared already in the early eighteenth century. The books by l'Hôpital (1696) or Maria Gaetana Agnesi (1748) using the Leibnizian form of the calculus, or the treatises by Charles Hayes (1704) or Humphry Ditton (1705) on fluxions tried to introduce the new technique early on. The defining treatises, however, would be the works of Leonhard Euler whose style and presentation would greatly influence the form of infinitesimal calculus in the latter half of the century (see Chapter 3).

The main genre of mathematical publications, however, were those used in the universities or in higher education that mainly contained arithmetic and geometry, though sometimes also elements of algebra. Although the contents were more in line with tradition, they did not escape the eighteenth century's preoccupation with reordering and resystematizing knowledge. Many people involved in the teaching of mathematics, often working at the peripheries of the world of the academies, wrote such books. These treatises allowed them to showcase their knowledge, both mathematical and historical, and their capability of using, structuring and reformatting existing mathematical themes. Their work appeared in publications of the local universities, in school programs or in their textbooks, and tried to get the attention of the local authorities so as to obtain a position or a pension. Topics popular among these authors included the status of negative numbers or of infinitely small numbers, and, to a certain extent, the parallel postulate. The development of these topics had a direct impact on the organization of a textbook or mathematics course. While these studies lay partly outside the academic realm, they influenced later nineteenth century discussions on how mathematics had to be organized and what should be its foundations and main elements (Schubring 2005).

Another example of this steady reworking of the systematics of mathematics was classical geometry. Although Euclid's *Elements* were often still the norm for the kind of mathematics taught in the universities, this did not mean that nothing changed in this more classical part. While the axiomatic system presented in Simon Grynaeus' edition of Euclid's *Elements* (1533) still prevailed in many geometry textbooks, since the seventeenth century the delicate architecture of definitions, axioms, postulates, and demonstrations had been in movement. In an eighteenth century obsessed with first principles and systems (see also Chapter 5), many explorations in rearranging Euclid were undertaken, many of which were presented as restoring Euclid's coherence from the extant texts and fragments (De Risi 2016). In the course of this rearranging, the parallel postulate increasingly became a contentious topic. The parallel postulate says that if a straight line intersects with two other straight lines, and has interior angles on the same side that add up to less than two right angles, the two straight lines, if extended indefinitely, meet on the side where the angles are less than two right angles.

This postulate did not seem to have the same evidence as Euclid's other postulates and axioms. What happens when the two interior angles come very close to 180°? Do the straight lines really intersect, or do they perhaps just get closer and closer? This question concerns, among other things, the concept of "straightness," which was not clearly defined by Euclid but assumed to be intuitively obvious. Mathematicians such as John Wallis or Girolamo Saccheri had proposed alternatives. In 1763, Georg Simon Klügel collected over thirty attempts to prove the postulate in his dissertation. Klügel's work stimulated Lambert in 1766 to try to demonstrate the invalidity of the cases where the angle sum is smaller or greater by logically deducing contradictions from these hypotheses, an approach already attempted by Saccheri. Lambert could not derive a contradiction, but by analyzing the geometry in algebraic form, he discovered that the sum of the angles of the triangles was proportional to their area and that an absolute measure of length existed. For the first case (sum of angles greater than 360°), Lambert gave an interpretation with triangles on a spherical surface. For the second case (sum of angles smaller than 360°) he postulated an "imaginary spherical surface," today called a "hypersphere," considering a sphere with an imaginary radius ir (with $i = \sqrt{-1}$) His results anticipated important parts of so-called non-Euclidean geometries, which were developed in the early nineteenth century by Gauss, Johann Bolyai, and Nicolai Lobachevsky (see also Chapter 3 in Volume 5).

MATHEMATICS AND MATHEMATICIANS IN PUBLIC OPINION

Even though religion was no longer a driving force behind armed conflicts, it was still very much present in society. The arena of debate, however, had shifted from physical confrontation to the Republic of Letters. For the greater part of the eighteenth century, books on religion made up for about a third of overall literary production, and many heated discussions in journals and correspondence turned around issues of divinity. Even Isaac Newton devoted an enormous amount of his time developing schemes to synchronize the Bible with historical evidence in his *Chronology of Ancient Kingdoms Amended* (published 1728). While his work in mathematics and rational mechanics had made Newton one of the torchbearers of Enlightenment, his *Chronology* uses instead the methods of early modern scholarship (Buchwald and Feingold 2013). This scholarship, a bookish erudition relying on comparing textual testimony, would slowly be replaced in the eighteenth century by a philological approach that used external disciplines, such as chronology, linguistics, numismatics, archeology, etc., to interpret their sources. The majority of theological debates would happen exactly on this battleground of interpreting the Holy Bible (and the world as a creation of God) dogmatically or dynamically, by bringing in other fields of knowledge.

Mathematics had a humble though steady part in this Republic of Letters. As it played a part in providing the theoretical backbone of influential worldviews, it did not escape entanglement in debates on natural philosophy and theology. Mathematics was crucial in physico-theological systems and many worldview systems, Cartesian, Newtonian, or other, hinged upon an interpretation of mathematical machinery. Voltaire singled out Newton's *Principia* as a masterpiece among humanity's achievements, all while using it to support deism, which illustrates the place of mathematics for (radical) Enlightenment thinkers and their opponents (see Chapter 4).

Although mathematics had a respected status, it was often ridiculed in the printing press. A botched eclipse prediction from 1699, the financial Bubbles of 1720, the many unusable methods of finding longitude or in general the lofty talk of those who talked mathematics but were inexperienced in practical matters had eroded popular esteem for mathematics. The tension between the practical man and the professional of mathematics was surely echoed here. In Jonathan Swift's book *Gulliver's Travels* (1726–7) the mathematicians of Laputa were tellingly depicted as alienated from their own world, though pretentious enough to want to govern the world:

> although they are dexterous enough upon a piece of paper, in the management of the rule, the pencil, and the divider, yet in the common actions and behaviour of life, I have not seen a more clumsy, awkward, and unhandy people, nor so slow and perplexed in their conceptions upon all other subjects, except those of mathematics and music. They are very bad reasoners, and vehemently given to opposition. [. . .] perpetually inquiring into public affairs, giving their judgments in matters of state, and passionately disputing every inch of a party opinion. [. . .] I rather take this quality to spring from a very common infirmity of human nature, inclining us to be most curious and conceited in matters where we have least concern, and for which we are least adapted by study or nature. (Swift 1726: pt. 2, 27–29)

This image of the mathematician, or of the scientist in general, as a stranger to this world though meddling with everything, seems to be a child of Enlightenment, too. Its most common depiction is that of an astronomer so deeply engaged in watching the stars that he falls in a pit. One mathematician around whom many images, tales and opinions crystallized was Isaac Newton. Already during his lifetime, Newton had cultivated and nourished his public image, but after his death the images multiplied (Fara 2003). His heroic status was already announced by Alexander Pope's epitaph, "Nature, and Nature's Laws lay hid in Night, but God said, Let Newton be! and All was Light" (1735). This reputation was cemented further by the cult of genius that became popular in the late eighteenth century. In the exaltation of originality over imitation and emulation, the more famous actors of what later would be called the Scientific Revolution were enlisted as geniuses among writers and philosophers: Johannes Kepler, Galileo Galilei, Nicolaus Copernicus, Francis Bacon, Robert Boyle, and Newton.

Indeed, as during the eighteenth century, the public sphere of discussion developed further, the scientist as such also became a public figure. The spectacle of science, be it through public shows of instruments and experiments of physics, astronomy, or electricity, or through displays and discussions in the salons and cabinets, had made science available to public opinion. The figure of the scientist now also slowly acquired some contours. Newton was definitely one of the molds, but many other types existed. The blind mathematician Saunderson (and later the blind Euler) was popular in discussions among philosophers as to whether those who did not have access to all five senses could develop reason. Johann Kasper Lavater's physiognomy (1775), the

FIGURE 0.5. Helluones librorum or the Bookworms by Joshua Kirby Baldrey (1786). The bookworm is reading Newton's *Principia mathematica* and both the walls and the floors are littered with mathematical literature. Photo by Heritage Art, Heritage Images / Getty Images.

derivation of character from facial profile that became very popular in the late eighteenth century, had as one of its early inspirations the facial countenance of the polymath Johann Heinrich Lambert. Lavater's collection of outline profiles contained the profile of an astronomer and mathematician, the most rational and noble character "even on the day all silver was robbed from his house."

As participants in the discussions on enlightened matters, such as chasing superstitions away and exposing theological bigotry, astronomers and mathematicians also made their mark. When Christian Wolff tried to show that the Chinese had a moral system (that of Confucius) of equal complexity to that of Christianity, he was fired from the University of Halle (1723). Charles-Marie de La Condamine publicly promoted the variolation method, imported from the Ottoman Empire, to counter the spread of the smallpox, though his success remained restricted to the aristocratic class (1755–73). The astronomer Jerôme Lalande was notorious for publicly showcasing the eating of spiders and worms at salon gatherings, thereby wanting to stamp out the received opinion that spiders were poisonous. Lalande was one of many astronomers and mathematicians that were suspected of atheism or deism at best. Lalande himself traced his beliefs back to the philosophical circle around Frederick II the Great that was reputed to be filled with atheists and deists, including such men as Voltaire and Francesco Algarotti, both staunch promoters of Newton's theories. In the *Dictionnaire des athées* (Dictionary of the Atheists, 1800), to which he contributed, Lalande stated:

> The spectacle of the heavens seems for everybody a proof of the existence of God. Coeli enarrant gloriam Dei. I believed this when I was 19 years old [when Lalande visited Potsdam and the circle around Frederick], today I see only matter and movement. One often asks me: but you that observe the sun, the moon and the stars, how can you not see the Supreme Being? I answer: I see there is a sun, a moon and stars, and that you are an animal. (Lalande 1803: 11)

Lalande's words announced the famous, though possibly apocryphal, anecdote attributed to Pierre-Simon de Laplace. When Laplace brought the final volume of his *Mecanique Céleste* (1805) to Napoleon, the latter asked where God was in his theory, Laplace allegedly replied: "Sire, je n'ai pas eu besoin de cette hypothèse" (Sire, I did not need this hypothesis). The instrumentalization of mathematics for causes against superstition and religion also made mathematics a favorite target of the opposite parties. Poets like William Blake or the German Romanticists chose to denounce mathematics as the sole instrument of reason, criticizing that it ossified knowledge into mere empty signs, all while appropriating mathematics' creative potential for literary use (see Chapter 7).

THE QUANTIFYING SPIRIT AND NATURAL PHENOMENA

In many respects, the Age of Enlightenment started the scientific and societal trend towards quantification of all kind of phenomena. This "quantifying spirit" (Frangsmyr, Heilbron, and Rider 1990) was the result of a number of evolutions. Philosophically, the revival of corpuscular and atomist philosophies in the seventeenth century was crucial for underpinning a quantitative and mathematical reading of the world. Technologically, the many advances in instrumentation enabled and harnessed a quantitative approach to a gradually enlarging range of phenomena. On the level of communications, the global access to the world combined with a reliable way of communicating through letters made the collection of data about many phenomena possible on a scale hitherto unseen. Finally, the mathematics of symbolic algebra and infinitesimal calculus provided powerful and very general ways of attacking a number of problems, while newer branches of mathematics, such as probability, algebraic geometry, differential equations, etc. made new fields of knowledge amenable to mathematization. Moreover, the success of mathematics, including the rather mundane use of arithmetic in trade and commerce, facilitated its application to new fields.

In the history of science, traditionally, the unifying mathematization of mechanics has been seen as one of the most impressive and lasting achievements of the eighteenth century. This mathematization, mostly associated with Newton's *Principia Mathematica* of 1687, brought both the celestial mechanics of the planets, satellites and comets, and the terrestrial movements of bodies in motion and collision together in one coherent mathematical framework. However, the development of rational mechanics was much more complex. It integrated both Newton's and Leibniz's ideas and techniques, and was gradually extended by Johann and Jacob Bernoulli, Leonhard Euler and many others. Through Euler's work and Joseph-Louis Lagrange's later synthesis, rational mechanics became a thoroughly mathematicized and powerful theory by the end of the century (Chapter 5).

For many parts of this rational mechanics, it proved to be quite challenging to derive the right mathematical equations and compute actual solutions. Questions such as the shape of the earth, the orbit of the moon and of comets, the three-body problem and many others attracted direct theoretical attacks with the new methods of fluxions or of differentials, but the differential equations that were derived could often only be solved by numerical approximation. This also meant that the computed approximate results depended heavily on the empirically determined constants of the equations. Every imprecision in these empirical values was propagated throughout the computations, and thus, the computed results did not always conform to what was being observed. The perfecting of rational mechanics therefore went hand in hand with a general improvement and perfection of measuring methods and instruments and with an increase in the number of measurements and observations. Lenses were improved, micrometers and more accurate measurement divisions allowed more precise observation, and techniques and technologies were developed to minimize or eliminate systematic instrument errors. For example, John Dollond's achromatic lenses allowed the reduction of chromatic aberration in astronomical observation, while Tobias Mayer's reflecting circle made it possible to diminish graduation errors in scales of measurement. As measurements became more precise, so did the numerical approximations to solutions.

Determining the shape of the earth, as the earth was not a theoretically perfect rigid body but rather a fluid ellipsoid, was a difficult and intensely debated topic, setting the Cassini family, astronomers at the Paris observatory,

FIGURE 0.6. The Observatory of Paris in 1740, with people observing Saturn and the moon. Photo by Oxford Science Archive / Print Collector / Getty Images.

against other members of the Paris Academy. Only when the Academy sponsored the Lapland and Peru expeditions, to gather empirical observations on the lengths of the meridian arcs, did the hypothesis that the earth was flattened at the poles win out (see Chapter 5). The theory of the moon, as a special instance of the three-body problem, was equally delicate and of pivotal importance for computing lunar tables that could be used to find longitude at sea. While theoretical mathematicians such as Euler provided lunar theories, it was the tables that Tobias Mayer computed on the basis of many observations that provided enough empirical accuracy for their use in practice (even if they were scarcely used at sea, see Chapter 6).

The theory of comets, especially those with elliptic orbits, was also taken up by many mathematicians, all producing different theories with different simplifications usable under different circumstances. The return of the comet of 1682 (Halley's Comet) was successfully predicted for 1759 by a number of astronomers (Nicolas Louis de Lacaille, James Bradley, Alexandre Guy Pingré, Alexis Clairaut) and bolstered support for Newton's system. In 1769, Charles Messier observed a new, very bright comet, and the computation of its elliptic orbit occupied astronomers for years to come. Also, the perturbative effects of the two big planets, Jupiter and Saturn, on their respective orbits and on the orbit of the earth and moon came to be studied. Lambert made a graphical analysis of the observations of their mutual resonances, while Laplace later

(1785) derived their periodic inequality analytically. With the improved observation instruments of the late 1700s, many new objects and phenomena were discovered, and new or improved theories had to be created. William Herschel discovered Uranus in 1781 though it took some years before it was unequivocally recognized the celestial object was not a comet, but a planet. In 1801, Giuseppe Piazzi observed Ceres, a dwarf planet, and again, astronomers began competing to compute the most accurate orbit. Eventually, the young Carl Friedrich Gauss' computations won out over Johann Carl Burckhardt's computations at the Paris observatory. This was made possible not only by Gauss' masterly computational skills and the use of a least squares error minimization, but also by the steady flow of new information and data that the astronomer-organizer Franz Xaver von Zach had steered in Gauss' direction.

Improvements in observation techniques were matched by an increasing internationalization of astronomy. Joseph-Nicolas Delisle and de Lacaille both actively stimulated astronomers around the globe, whether academicians, sailors, observers at colonial outposts or Jesuits in India and China, to observe the same objects or phenomena simultaneously, such as the solar eclipse of 1748, the conjunction of the moon, Mars and Venus in 1751–2 or the transit of Venus of 1761 and 1769. The transit of 1761 was observed from over 100 locations by nearly 200 different people. One of the objectives of these observations in parallel was to determine more exactly the parallax of the sun, that is, the differences in apparent position of the sun as seen from different places. De Lacaille also observed stars at the Cape of Good Hope, and drew up a catalogue of stars (1763) that widely surpassed the previous atlases of Johannes Hevelius and John Flamsteed, both in the range and in the precision of observed stars. This worldwide network of observation posts was initially connected through personal correspondence and, later, collected in publications in the Academy memoirs. Near the end of the century, the first astronomy journals would appear. Under Lambert's guidance at first, and through Johann Elert Bode's work later, the *Ephemeriden* of the Berlin Academy were converted into a more scientific periodical. Another important catalyst behind internationalization was Franz-Xaver von Zach whose correspondence and journals collected and regularly printed observations from all over Europe.

Though accounting in computational detail for every celestial phenomenon remained a challenge, it became increasingly plausible that the mathematical model for the stars, planets and other heavenly objects could, at least in theory, provide that detail. This idea was worked out and popularized in a number of influential books. In 1685–6 Bernard de Fontelle wrote his *Entretiens sur la pluralité des mondes,* that caused upheaval mainly because of its speculation of living creatures on other planets. In 1761, Lambert published his *Cosmologische Briefe*, introducing probabilistic arguments to account for the distribution of the stars and giving an explanation for the Milky Way (independently proposed by Thomas Wright in 1750). Finally, in 1796, Laplace published his *Exposition du Système du Monde*, a universe where calculation reigned supreme and culminated in the demonstration that the solar system was stable, meaning that it would not degenerate into chaos. Laplace's multi-volume *Mécanique Céleste* was supposed to provide the elaborate computational proof.

The quantification of mechanics also tackled more earthly matters such as the movement of cannon balls, human beings and machines, or the movement of the waters, an important topic in the eighteenth century. The new calculus was put to work, but it was confronted with reality checks. Without empirical data and without a sense of what could be neglected in practice, the analytical models did often not produce any useful results. Experimentation in the field and the development of new or more accurate measuring instruments were crucial for advancing this program. Building upon existing know-how in the instrument workshops, new instruments were devised, or old designs much improved.

In ballistics, Robins' ballistic pendulum enabled the measurement of a bullet's impact. His results would spur Euler to rework his theoretical approach (see Chapter 6). Especially in and around the military academies near the end of the eighteenth century, practical experiments were conducted and slowly matched up with theoretical treatments, improving the computation of ballistic tables (see Chapter 2). In the movement of waters, Daniel Bernoulli published his famous *Hydrodynamica* in 1738, spelling out a framework that applied calculus to hydraulic problems (see Chapter 3). In practice, however, the management of rivers and canals, or the improvement of water mills, had to wait until the instrumentation improved to do experimental measurements, with devices such as the hydraulic speedometer, the palmette wheel, the Pitot tube, etc. The establishment of institutes in Italy and France to systematically measure and study water movement near the end of the century laid the groundwork for the parallel development of theory and practice that was eventually highly successful in the nineteenth century.

The successful mathematization of mechanics opened possibilities for doing the same for other natural phenomena. In the more experimental sciences, which we may call "Baconian" sciences (Kuhn 1976), the parallel development of measuring devices and theoretical models rendered them more and more quantitative. New instruments were in continuous development, making a wide variety of natural phenomena amenable to quantification. Examples include thermometers to measure heat, hygrometers to measure humidity, and torsion balances to quantify electrical phenomena (see also Chapter 5).

The instrumentation of chemistry also helped chemistry evolve into a quantitative discipline. For the philosopher Immanuel Kant, in his *Metaphysische Anfangsgründe der Naturwissenschaft* (Metaphysical Foundations of Natural Science, 1786), chemistry was still intermediate: a rational science, but not a proper, mathematized science. However, at the time of Kant's writing, quantification was already winning its ground in chemistry with Joseph Black's analytical balance and chemical oven, with Antoine Lavoisier's precision gasometer or, later still, with Jöns Jakob Berzelius' electrolyte experiments. At the same time, numerical and algebraic ways of expressing chemical reactions gained acceptance, such as, e.g., Jeremias Benjamin Richter's stoichiometry or Berzelius' chemical formulae.

Another field in evolution was the more experimental, earth-oriented brother of astronomy: meteorology. With instruments to measure temperature, humidity, and air pressure, many learned men started to systematically collect data about the weather. Leiden professor and influential Newtonian Petrus van Musschenbroek was one of the first to publish his findings, introducing a new format of displaying numerical data along the way. Instead of putting his data into tables, Musschenbroek visualized his data as dots, their vertical position indicating the numerical value of temperature measured, their horizontal

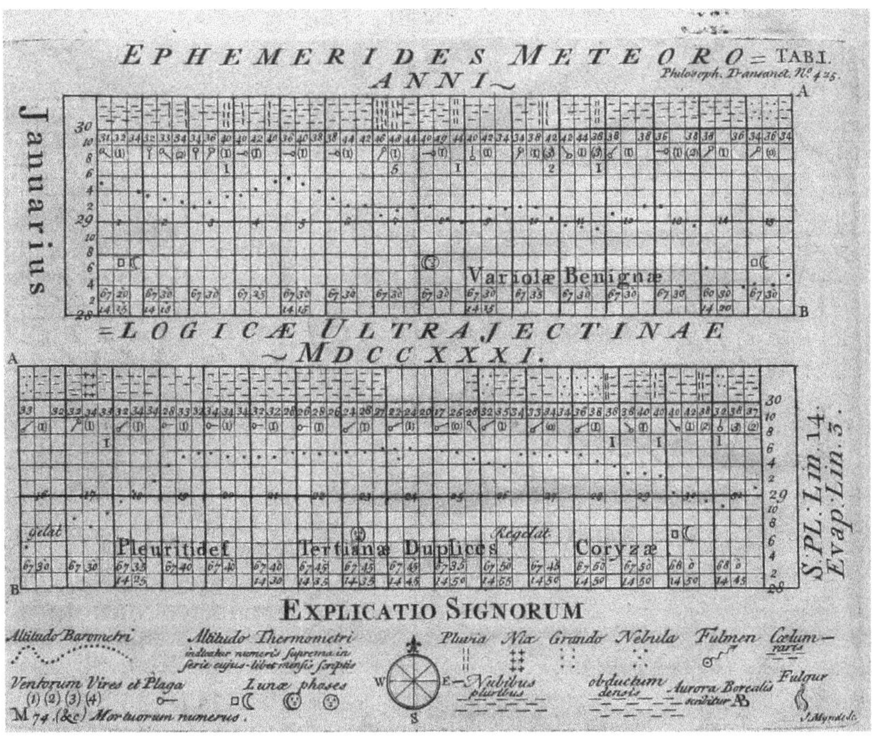

FIGURE 0.7. Musschenbroek's visualization of meteorological data for 1731. The top level indicates the kind of weather (rain, clouds, misty, sunny), the middle level gives the temperatures as dots, though they are not as yet connected by lines. Photo courtesy of The Royal Society / Internet Archive.

position bound to the time of the year. He also advocated a world-wide network of observatories that would record meteorological data around the world. Musschenbroek's early visualizations would be taken up by other early meteorologists. In the hands of Johann Heinrich Lambert, however, the visualizations would be combined with mathematical analysis. Lambert zoomed in on parts of the graph, derived new graphs geometrically and mechanically, tried to find its characteristic symptoms and asymptotes so as to find analytical expressions that matched the empirical curves. The visualizations were quite literally used as a bridge between the "Baconian" parts of science and its more quantitative parts (Bullynck 2010b). The use of visualizations of data in combination with mathematical analysis foreshadowed here, came into more general use in the nineteenth century, in particular in combination with statistical methods.

THE QUANTIFYING SPIRIT AND SOCIETY

The quantifying spirit of the eighteenth century did not stop at the phenomena of the natural world, but spread out to various domains of human society. The growth of worldwide trade as an intricate system of assets and profits, risks and losses underwrote an economic view on society, and as such it also contributed considerably to the expression of the quantifying spirit of the eighteenth century. The double-entry bookkeeping of the Italian merchants, the detailed ledgers of the banks and shipping companies or the voluminous treatises on monetary exchange and trading law all attest to the convoluted techniques of doing commercial arithmetic that had been developed in Modern Europe. While, intrinsically, the mathematical content did not go beyond quite elementary arithmetic, the sheer diversity of rules and the combinatorics it created made commercial arithmetic quite daunting. The title of Joseph de la Vega's book describing the Amsterdam stock exchange, *Confusion de confusiones* (1688), is quite evocative of this complexity.

Teaching commercial arithmetic was a trade in its own right. The lessons were privately offered and would introduce its students to the world of calculating with fractions, of computing exchange rates and of drawing up account books, inventory ledgers, or contracts. Those topics went beyond the elementary arithmetic on offer in the corner schools and grammar schools, where fractions were rarely taught. By the end of the eighteenth century, however, commercial arithmetic became institutionalized in the more advanced special schools that educational entrepreneurs or some states opened in the late eighteenth century (see Chapter 1).

The mathematically most complex products in the commercial world were certainly the insurance firms and the moneylending business, where chances and expectancies had to be taken into account. It is no wonder that exactly in the centers of banking and insurance, such as Amsterdam, London, or Geneva, the interest in the mathematics of compound interests, life annuities, and so on was most articulated. On the basis of empirically constructed mortality tables and with the use of probabilities, various theories of life expectancy and tables for annuities were computed (see Chapter 4).

A rather indirect effect of these intricate financial constructions was that they made a number of properties of the human world subject to quantification. In particular, life and death had become probabilistic events. John Graunt's book, *Natural and Political Observations Made Upon the Bills of Mortality* (1662), was one of the first to derive averages and life expectancy tables from the empirical mortality tables. Jacob Bernoulli later took up the idea and started assigning numerical values of expectancy to events so as to solve a number of societal problems mathematically. These numerical values were obtained by summing up the cases that had been empirically observed or collected, divided by the number of cases (the arithmetic mean of the data).

This evolution did not happen without much (philosophical) discussion. The concept of "chance" in the eighteenth century was intimately connected with the understanding of the world as (mechanistic) system where causes led to effects, possibly governed by a deity. Chance only had a place in the "voids" of this world of causes and effects. To some, the calculus of probabilities could thus only be applied to effects not having happened yet and where the chain from cause to effect, and thus the range of possible effects, was known. The classic games of chance were examples of this. To others, the human incapacity to perceive the causal chains in many phenomena allowed for the use of probabilities in other problems, too, using numerical values that were empirically obtained.

This was one of the crucial points setting d'Alembert against Daniel Bernoulli, who had refined and applied his uncle Jacob's idea. D'Alembert's critique of probabilities boiled down to the more conservative viewpoint:

> if one computes something using the laws of chance, it is always about an effect that has not happened yet; and as all effects are equally possible, one can easily determine how much to bet upon an effect not happening. But once an effect has happened, it no longer makes sense to use the laws of chance. (article "*Inclinaison*" in the *Encyclopédie*)

Applied to the discussion on Daniel Bernoulli's mathematical theory of inoculation (1760, published 1765), this meant for d'Alembert that for the individual, the possible death incurred by being inoculated rendered the probabilistic computation for or against inoculation void. For Daniel Bernoulli, on the contrary, individual trajectories did not matter, only whether, on average, life expectancy increased or not if persons were inoculated.

This approach was pioneered by the Bernoulli family, and they applied it to questions in medicine (disease, inoculation), law (inheritances), to questions of morality and astronomy, and to logical propositions. Evidently, its practical utility for insurances and annuities was not lost on the Bernoullis and their circle of correspondents in Switzerland and elsewhere. Abandoning d'Alembert's doubts about using probabilities in the study of societal problems, a newer generation of the mathematically adept in France also embraced the use of probabilities in moral and political questions. After the quantitative approach to governing advocated by Anne-Robert-Jacques Turgot, Nicolas marquis de Condorcet would enthusiastically propose a political arithmetic. This discipline would not only gather facts and/or opinions, but also evaluate them and derive mean values, and, finally, follow their consequences and compare their probabilities. By this means, the most probable or best system could be found.

The quantification of society was not pursued only in the financial and commercial world. Governments, small or big, became more and more invested in what came to be called "statistics," the general overview of resources of a state or town, especially regarding its territory, its population, its natural resources and its industry. Not a small part of this was related to things financial. With the complicated financial constructions of the eighteenth century, cities and states had to gain a "rational" overview of their income and resources to be able to either take up new loans or redress their debts. The main question was how to set up a stable and profitable tax system. Much discussion focused on what tax systems would be best: either taxing produced goods such as beer and wine, or agricultural produce, or rather raising money through a tax on land and property? Detailed and reliable statistics of the state were duly needed.

It is not surprising, then, that detailed maps of the land were among the first things need for statistical description. Some empires, such as France and Russia, had a central organization for achieving an overall map, while others, such as the United Kingdom or Prussia, relied on standardization of surveying practices to patch together a global map of the land (see Chapter 2). Hand in hand with the mapping of the territory, censuses of goods, resources, and population were undertaken. Increasingly, the number of births, marriages, and deaths as found in the registers of the church or of the cities were taken down, put in tables, and analyzed. Discussion remained whether mathematical, say statistical, techniques were needed for their analysis and use, or whether the significance lay rather in the exhaustiveness of the description. The many lists and tables that were compiled were not only of interest to the administration of the cities or states, but also to the learned and to a general readership. Moreover, they figured prominently in the first "travel guides" appearing in the later eighteenth century and addressed a growing mobile class that could undertake leisure travel along the postal routes.

With the new infinitesimal calculus, the question of finding the maximum and minimum of a curve or of a family of curves had become an important focus of research, laying the basis for what later came to be called the calculus of variations (see Chapter 3). The idea was repackaged by Maupertuis in a more appealing and philosophically more accessible way as the principle of least action (1751). It seemed that the extremal values (maximum or minimum) coincided with the preferences of nature's course and offered a mathematical way of studying nature and the world. Though most studies concentrated on mechanics (see previous section and Chapter 5), other domains were also attacked. A famous example was a challenge that the physician René-Antoine Réaumur had formulated, namely, to prove that the beehive's hexagonal structure was an optimal solution of building with restricted resources. In 1739,

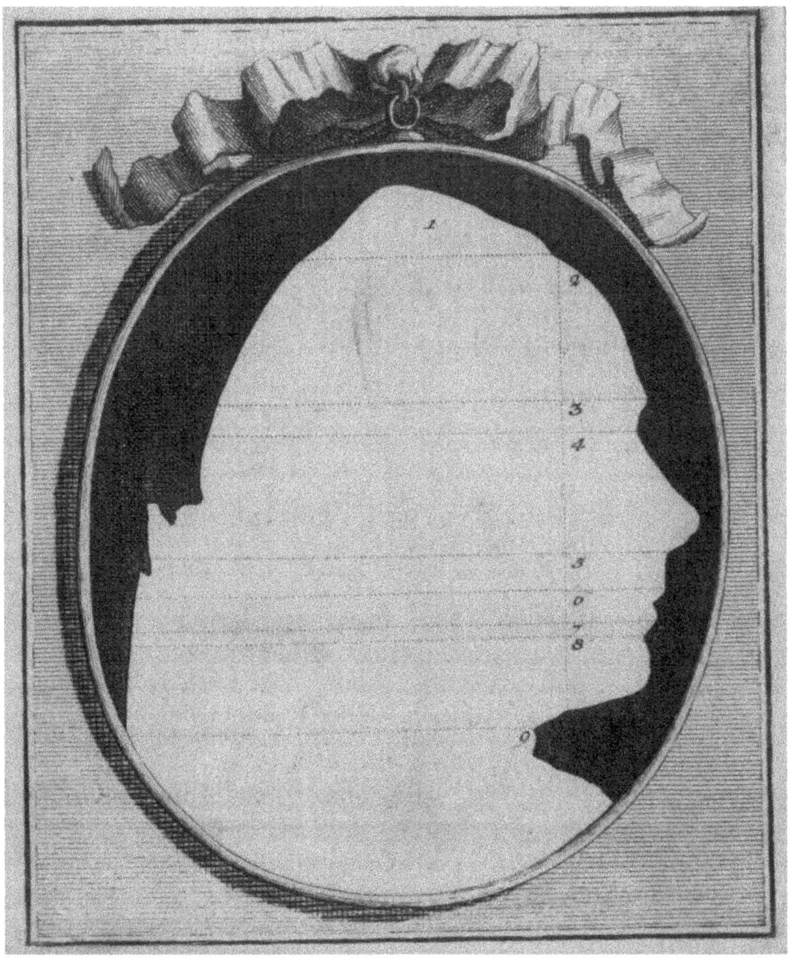

FIGURE 0.8. A silhouette with a grid from Lavater's *Physiognomische Fragmente*, the part between 2 and 3 is described as a parabola. Photo courtesy of ETH-Bibliothek Zürich, Rar 9206, Public Domain.

Samuel König of the Berlin Academy mathematically demonstrated that a hexagon was indeed optimal. This kind of reasoning, though often rather informal rather than with mathematical precision, surfaced often in eighteenth-century treatises, especially in those authored by Christian Wolff. The principle not only appeared in nature, where it governed equilibrium and movement alike, but it was also expressed in human constructions, such as building, paintings, literature, etc. William Hogarth's *Analysis of Beauty* (1753) is similarly concerned with giving a rational account of art and beauty. Hogarth discerned elegance and fine proportion in the agility of the movements between two extremes, much as the extremal principles in mechanics found the way of nature's own course. To Hogarth, "the eye generally gives its assent to such space and distances as have been first measured by the feeling, or otherwise calculated in the mind." The same ideas about optimality also offered a way of quantifying aesthetic categories. Johann Heinrich Lambert, for instance, devised a calculus to find the most aesthetically pleasing arrangement of things (1769), which amounted to finding the minimum of displacements among things. Even Lavater in his *Physiognomische Fragmente* (1775) suggested that mathematical curves could be recognized in detailed silhouettes of people's faces, "it is to be hoped that a mathematical genius—will enter this course of study and will apply his powers to the curves of humanity."

REVOLUTIONS

On the whole, the eighteenth century was characterized by growing tensions between modernization efforts and systematic resistance against modernization by the entitled and privileged parts of society. These tensions flared up chronically but never really exploded, until 1789's French Revolution, when the century goes out with a "bang." While, from the historians' point of view, the French Revolution was a long time in the making and certainly the confluence of many slow development and critical events, it nevertheless had an enormous impact on contemporary society and an important afterlife in how history was written and remembered. Symbolically, the beheading of Louis XVI (1793) sent shock waves through Europe and the world. Practically, the wars of the early nineteenth century had a more lasting impact on changing the everyday structure of society. The Revolutionary Wars (1792–1802) and the Napoleonic Wars (1803–15) overturned many of the existing power structures in Continental Europe and prepared the ground for the modernization of the nineteenth-century states in many respects. They likewise revolutionized colonial relationships. The Haitian Revolution (1791–1804), despite Napoleon's invasion to keep France's most profitable colonial asset in French hands, led to an independent Haiti. Seeing that it was impossible to pursue war both in North

America and in Europe, and needing money to finance the wars, Napoleon initiated the Louisiana Purchase (1803) selling the Louisiana territory and the preemptive rights to acquire native American land by treaty or conquest to the young United States of America. After Spain's and Portugal's loss of power during the Napoleonic Wars, many of the Latin American countries went for independence, too. On the other hand, England was able to expand its colonial empire and France made its first incursions into Africa with Napoleon's expedition to Egypt.

The French Revolution and its aftermath also became a symbol for the new, central role science and mathematics came to occupy in modern society. The merits of the sciences and of mathematics for organizing society more rationally had been extolled throughout the eighteenth century, and though at various moments in time some ideas were cautiously introduced or attempted, the promise was never actually realized. Under the First French Republic (1792–1804), the beheading of Louis XVI, the subordination of the Church to the Republic, and the attempted dechristianization had literally removed the ancient symbols of state and church. In their place, a cult of Reason and other rationalized forms of religion were promoted. The sciences and mathematics, as the torch bearers of Reason (a stock image of the Enlightenment), were duly enlisted to supported the new state religion.

As part of this restructuring of society, the decadic system was introduced everywhere. Of course, for some centuries, the Hindu-Arabic decimal positional system of writing numbers and doing arithmetic had been in use. However, many domains of practical everyday knowledge were not base 10. Under the First Republic, the decimal system was generalized to weights, coins and measures (the so-called Metric System, see below), to get rid of the local variety of weights and measures of the Ancien Régime. Divisions of time were also amended to fit a Republican Calendar and get rid of the Gregorian Calendar. The calendar started anew from year 1, An 1 de la République française, on September 22 1792. The year counted 12 months of 30 days, the seven-day week was replaced by *décades* (a ten-day week), and the day had ten hours, each of 100 decimal minutes (of 100 decimal seconds). As a result, new measuring instruments were needed, new clocks, new calendars, and also new tables of trigonometric functions as the circle now was subdivided in 400 decimal degrees. Though the metric system would ultimately, after some setbacks, become a success and be generally adopted in Continental Europe, the Republican Calendar was abolished on December 31 of the year 1805. Nevertheless, it testifies eloquently both to the determined will to change things profoundly, and to the exemplary value of mathematics during the Republican period.

Many persons well versed in mathematics took on important functions in this First Republic. Nicolas de Condorcet, who had long advocated that the

FIGURE 0.9. The Republican Calendar. Photo by Fine Art Images / Heritage Images / Getty Images.

political and moral sciences might follow the example of mathematics, was member of the governing Convention Nationale in 1792 proposing a reform of the educational system, though he was arrested and guillotined in 1794. Antoine Lavoisier, the famous chemist and notable mathematician, undertook a computation of the economic needs and resources of France together with Joseph-Louis Lagrange (1791), but, due to his function as a tax collector (as *fermier général*) under the Ancien Régime, was guillotined in 1794. Lazare Carnot was also a member of the Convention and part of its successor the Directoire. As a general in the army he was an important factor in the Revolution's military successes. Pierre-Simon de Laplace, Joseph Fourier, and many others held public functions during the First Republic.

The two most important scientific ventures of the First Republic with a lasting legacy belonging to the history of mathematics were the reform of the (higher) educational system with the foundation of the École polytechnique and the Écoles normales (see Chapter 2 and Volume 5), and the appropriation of a project dating back to the Academy of the Ancien Régime, namely the establishment of a universal system of weights and measures, a project that would become the Metric System.

The idea of a universal system of weights and measures had long been the object of discussion among scientists. The description of nature, in particular of the earth and the solar system, had become an important endeavor with both scientific and imperial importance. In science, it helped assess and develop parts of Newton's theory of gravitation, for the imperial states, knowledge of the earth and the stars became ever more important for drafting maps and improving navigation (cf. Chapter 2). Therefore, there was a need for scientific measurements to be shared not only nationally, but also internationally. The germ of what later would become the (Continental) metric system was originally planned as a joint scientific enterprise of British and French astronomers. The plan would have been a follow-up to the agreement joining the Greenwich and Paris observatories in 1789 that made the French and British measurement grids compatible. The French Revolution and the installation of the first Republican government (1792) killed off that joint enterprise. The new French government, however, seized upon the occasion and turned an essentially scientific project into a general reform of the weights and measures of the Ancien Régime.

The idea of the metric system is built upon the logic of the decimal positional system for numbers. One can consider decimal fractions as an extension of the decimal notation that substitutes the intricacies of manipulating ordinary fractions with a homogeneous system of decimal computation. Similarly, the metric decimal system is supposed to do away with the variety and complexity of weights and measures and introduce one common system of measures having the same structure and logic of the decimal system, where the smaller and larger measures have proportions that are powers of ten. The reference point around which the other measures and weights were to be defined was the metre, fixed (and experimentally observed) as the ten-millionth part of a quarter meridian of the earth.

After the scientific committees had done their work, the metric system was rendered obligatory for the entire population of the French Republic by law on the 13 brumaire An IX of the Republican calendar (corresponding to November 4, 1800). The propaganda machine of the Republic put its force behind the dissemination of the metric system, putting the new standard measure on public display and hanging up placards explaining the new system. It also made the metric system a central part of arithmetic education, its teaching became obligatory and handbooks had to contain a section on it. This did not prevent the new system from meeting with much resistance among ordinary people. Not only do old habits die hard, but in a time of revolution yet another familiar aspect of the everyday world was threatened.

The aversion to the new system would finally translate into political decision-making. In 1812, Napoleon ruled that the old measures could be used once more in commerce and trade and that calculations with ordinary fractions were allowed again. With the Restoration of the monarchy in 1816, the metric system was even suppressed altogether. But it nevertheless slowly took over. In Continental Europe, the Netherlands was the first state to legally introduce it in 1816, other states followed, and France, finally, a generation after its first introduction, returned to the metric system under Louis-Philippe (1840). Great Britain and most of its (former) colonies, the United States of America, and Russia, which had not been under Napoleonic rule or influence, did not adopt the metric system and kept to their imperial system(s) of measures and weights.

CHAPTER ONE

Everyday Numeracy

The Slow March towards the Democratization of Numeracy

MAARTEN BULLYNCK

There was no such thing as a general form of numeracy in the eighteenth century, but rather a multitude of numeracies, stratified along social or professional classes and conditioned by regional boundaries. Each such numeracy had its own devices, notions, and rules, and was deeply embedded, if not inherited, in a practical context from which it could not be separated. In this respect, the eighteenth century is in many ways the continuation of previous centuries (cf. Chapter 1 in Volume 3). In the course of the eighteenth century, however, mathematics and in particular arithmetic, acquired more social prestige. Instead of being mostly associated with the clerical work of writing, copying, and accounting, its virtues for society as a whole and for improving the human mind began to be extolled. The trope of the utility of mathematics for the human mind and for mankind was not new, but gained momentum in the eighteenth century. This furthered the realization that numeracy, like literacy, should eventually become a cultural skill accessible to all people. In this process, the passage of many numeracies, each one tied to their peculiar practical context, to a general numeracy in the singular happened. The beginnings of numeracy as part of general (and compulsory) alphabetization are therefore to be found in the eighteenth century.

If one wants to have an idea of who was using arithmetic in the seventeenth and eighteenth centuries, one can turn to one of the most translated and reprinted books of the eighteenth century, Johann Amosz Comenius' *Orbis Pictus* (1657). This book is the first pictorial encyclopedia meant for both children and adults. Through 150 plates, it introduces the reader to all aspects of society and to the vocabulary to talk about it. In Plate CI, one encounters the many faces of arithmetic and reckoning:

> The Naturalist, 1. vieweth all the works of God in the World.
>
> The Supernaturalist, 2. searches out the Causes and Effects of things.
>
> The Arithmetician, reckoneth numbers, by adding, subtracting, multiplying and dividing; and that either by Cyphers, 3. on a Slate, or by Counters, 4. upon a Desk.
>
> Country people reckon, 5. with figures of tens, X. and figures of five, V. by twelves, fifteens, and threescores. (Comenius 1659: Plate CI)

The panorama goes from the lofty applications of arithmetic by learned philosophers and natural scientists, and stretches to the mundane uses of numbers and cyphering by reckoners and farmers. Numeracy could thus be found throughout the population. However, for each social or professional group, this

FIGURE 1.1. Image CI of J. A. Comenius' *Orbis Pictus* or *Visible World* (1657, English translation 1659), depicting the Metaphysician looking at the skies, the Physician looking at the ground, and, inside the house, the arithmetic practices of the reckoners and the farmers. Photo courtesy of Universitätsbibliothek Basel, UBH ki III 3. Public Domain.

numeracy had a different face, a different identity. The materiality (paper, slate, or desk), the notation (decimal ciphers, counters, or strokes), and also the amount of mathematics involved was different for each and every group.

But a certain percolation effect took place, the publicity around the lofty applications of mathematics trickled down to more earthly matters and preoccupations. The success of the mathematical method in the natural sciences made its mark on literature, philosophy, and theology and the usefulness of mathematics for science and society were quite widely promoted in Enlightenment literature. One of the first to expound mathematics' virtues for exercising the mind was John Locke in *Of the Conduct of the Understanding*:

> I have mentioned Mathematicks as a way to settle in the Mind an habit of Reasoning closely and in train ; not that I think it necessary that all Men should be deep Mathematicians, but that having got the way of Reasoning , which that study necessarily brings the Mind to, they might be able to transfer it to other parts of Knowledge as they shall have occasion. (Locke 1706a: 30)

Also the influential philosopher and manual writer Christian Wolff had praised mathematics both for its intellectual virtues and its practical utility. Baron Anne-Robert-Jacques Turgot, who had tried to reform the French tax system on a more data-based and scientific basis, had equally vaunted mathematics because it led to truth and avoided errors.

On the whole, this positive attitude towards mathematics, in combination with changing economic circumstances, made numeracy slowly appear more prominently on the political agenda, Indeed, both the internationalization of trade and the beginning industrialization created a new need for people conversant in (elementary) mathematics. Also, the state's role changed and in many places the court and administration evolved from being a class-based system to a mixed system where merit played an increasing role alongside class in the recruitment and advancement of civil servants (see Chapter 2). As the states tried slowly (and often unsuccessfully) to abolish old feudal privileges and install more modern ways of administrating, they needed civil servants that could readily apply a number of mathematical techniques. This would stimulate the authorities to organize some form of teaching both literacy and numeracy, supporting, supplementing, or even supplanting local or mostly individual initiatives. Although this was mostly limited to cities or regions that were tied in to international and/or industrial developments, this constituted a beginning of a more general democratization of numeracy.

If there is one watershed moment in the history of numeracy during the eighteenth century it would be the year 1773. That year witnessed the suppression of the Jesuit order and, as a consequence, the reorganization of the school systems in larger parts of Continental Europe (Bosse 2012). As the

Jesuits controlled much of the higher education in the catholic countries (see Chapters 1, 2, and 7 in Volume 3), their suppression left a vacuum and drastically opened up the educational landscape. For many states, this was the opportune moment to seize control of education and define its contents. This modernization of education gave a new élan, both institutionally and methodologically, to getting compulsory literacy realized. As part of this process, concrete steps towards establishing compulsory numeracy were spelled out too. Throughout the eighteenth century, many local initiatives had begun to spread numeracy, but only from 1773 onwards did a more general movement to establish compulsory numeracy grow and endure, even though it would take long into the nineteenth century until the ambitions of lawmakers and educationalists would be generally realized in practice.

While the horizon of eighteenth century numeracy is the inclusion of reckoning as elementary arithmetic into general, compulsory literacy, the landscape of numeracy throughout the eighteenth century remained rather of bewildering variety defying easy categorization and generalization. The geographical and social distribution of numeracy in the eighteenth century reflected a number of global societal divisions but mostly was the complex sum of local traditions and individual trajectories. As remarked in the beginning, quite some continuities to previous centuries persisted and many forms of numeracy coexisted in the longer eighteenth century.

THE (NOT SO) PECULIAR CASE OF CARL FRIEDRICH GAUSS

Rather than surveying the field, we will introduce this complexity by following a privileged witness who encountered many forms and modes of numeracy during his youth, Carl Friedrich Gauss (1777–1855). More even, these experiences helped to shape his becoming a mathematician. Gauss' biography features one of the best known school anecdotes in the history of science. As a school pupil, the young Gauss was given the exercise to sum a sequence of successive integers at his elementary school in his hometown Brunswick. Instead of laboring his way through many additions, he immediately wrote down the answer and with the words "Ligget se" ("Here it is"), put his slate with the correct answer on the master's table (von Waltershausen 1856). Although the anecdote is used to illustrate Gauss' precocious talent for mathematics, once framed in the broader context of late eighteenth century Brunswick, the anecdote rather reveals the social and institutional fabric mediating arithmetic in Gauss' time. It tells of the history of Gauss' family and of the evolution of Brunswick's educational system. Together, these provide the systematic background in which Gauss' talent was fostered and could blossom.

FIGURE 1.2. The young Carl Friedrich Gauss (1777–1855) in a gouache by Johann Christian August Schwartz (1803). Photo courtesy of Städtisches Museum Göttingen / Wikimedia Commons.

In Gauss' time, Brunswick (Braunschweig in German) was with about 26,000 inhabitants one of the larger German cities. It was an upcoming city, where the Enlightenment and modernization had set in during the eighteenth century. This was facilitated by the Dukes of Brunswick who had close political and economic ties with Hannover and England. As in many other cities of comparable scale that lived from their commercial relationships and industrial production, Brunswick counted a number of teachers, often called *Rechenmeister* or *Arithmetici*, that offered courses in arithmetic. Through payment of a fee, students could follow the teacher's courses, mostly at the teacher's home or in a rented space, the so-called "Winkelschulen" (corner schools). As a matter of fact, the elementary corner schools dispensed both training in writing and in reckoning, both being clerical, manual activities where signs had to be brought to paper or to slates. Writing and reckoning belonged together. In the cities and sometimes even in villages, schools would frequently host an arithmetics teacher, who mostly did not belong to the faculty of the school, but offered his services as an "extracurricular" activity, again against payment of a fee. Often, the more apt pupils would be enlisted by the master as helpers. The level of arithmetic taught was diverse. It could be only writing and reading numbers using the decimal positional system together with addition and subtraction, it could be the four species, or it could include advanced calculating rules such as the rule of three and many others. Moreover, it would often be a kind of "practical" arithmetic, a selection of elementary arithmetic and some calculating rules tailored to the needs of a specific trade or profession.

The situation in Brunswick testifies of a slowly transitioning towards a more state-organized educational offer and the will to disseminate numeracy more generally. Already in the sixteenth century, the city struck a deal with the church authorities that next to the schools that were backed by the church, other schools would also officially be supported by the city. These were two Latin schools, the Martineum and the Catharineum, and two *Schreib- und Rechenschulen* (schools to learn writing and reckoning). Their teachers received free accommodation and a flat fee, the rest of their income depended on the fees the pupils had to pay. Because of increasing competition with *Winkelschulen* in the eighteenth century, the two *Schreib- und Rechenschulen* obtained a ducal privilege that protected them against this competition. In a law text from 1755, Brunswick detailed how these schools should be run, what they had to teach and how it should be examined. It also determined how much the pupils had to pay. In the *Schreib- und Rechenschulen*, for example, 18 Mariengroschen for religious instruction and writing, 27 Mariengroschen for arithmetic (*Rechnen*) as well as 12 Mariengroschen contribution for heating the school room. (One may bear in mind that 3 Mariengroschen could buy you a big loaf of bread.) On top of that, for extra hours of arithmetic, additional fees were due (Koldewey 1898: 319–20). The state thus not only dictated the regulatory framework of all the schools, but increasingly determined what was taught and how it was taught, at least in those schools it partially funded. It should be noted that near the end of the eighteenth century, the schools supported by the city of Brunswick had also been opened up to girls. They made up about half of the number of pupils. The schools still had much autonomy though, also the state could not influence the private or church-backed schools in the same way. To give an impression of the educational landscape in Brunswick around 1784: it counted the 2 Latin schools, the 2 *Schreib- und Rechenschulen* as well as 2 other state-backed schools and

6 *Armenschulen* (free schools for the poor), next to 4 church-backed schools, and 37 *Winkelschulen* (Klehn 1999: 159).

In this process of increasing state intervention in education, Brunswick also pushed the teaching of numeracy. At the end of the seventeenth century, they had attracted a capable *Schreib- und Rechenmeister* (and *Buchhalter*, bookkeeper, or accountant), Johann Balthasar Remer from Lüneburg. Remer was member of the *Löbliche Arithmetische Societät* in Hamburg, one of the first mathematical societies consisting mostly of teachers and accountants. This society was a professional organization, but also fostered discussion and communication of mathematical problems, riddles, methods and solutions among its members. After Johann Balthasar's death, his son, Christian Stephan Remer, also member of the *Societät*, would follow in his father's footsteps and teach at the *Schreib- und Rechenschulen*. After Christian Stephan Remer's death, replacement was sought in in Brunswick but also in Hamburg, the job advertisement stipulating that those who could only write neatly and reproduce arithmetical examples should not apply, because they were looking for a more capable *Rechenmeister*, in particular for someone with skills in commercial arithmetic and bookkeeping. This school, the *Schreib- und Rechenschule* near St. Catherine's, is where the family Gauss would go to. First Gauss' father Gebhard Dietrich, a generation later, Carl Friedrich Gauss himself.

Both father and son Remer of the school would publish manuals, the father in 1690, the son in 1737. The respective content of their books and their introductions reflected the evolution in arithmetic education in Braunschweig. It showed how the city's continued efforts and support paid off as it produced a cumulative effect, the next generation being able to advance on the achievements of the former generation. Father Remer insisted on the fact that arithmetics was useful not only for merchants, but for all social classes. He thanked the government for supporting his schools, exhorting the parents to pay for their children to take classes in arithmetic and to avoid going to *Winkelschulen*.

His book was neatly split up in three parts: Book 1 gave the instructions, Book 2 the examples, and Book 3 the solutions. Further, the book contained the rules for the four species applied to integers and fractions, but also twenty more special rules for specific cases, such as money change, discount calculation, interest calculation or Italian practice (i.e., commercial arithmetic). These special rules could be acquired during extra, privately held tutoring against payment of an extra fee. The idea behind the three parts of the manual was that, after initial instruction, the teacher would let the pupil copy down the example, than let him go to his desk to work out the example and compute the solution before returning with the answer (or with questions on how to find the answer) to the teacher. This was conform to the structure of other *Rechenmeister*-books, such as the popular books written by Christian Pescheck from Zittau, and was reflexive of actual practice in the schools.

FIGURE 1.3. Frontispiece of Christian Pescheck's *Arithmetischer Haupt-Schlüssel* (1735). It depicts the Rechenmeister dealing out examples to the pupils, once the pupils find the solution, they have to return to the teacher to check on it. In the back is a cupboard with drawers on which are written the names of special rules. It is a metaphoric representation of Pescheck's book, where for each example, the right drawer, i.e. the right special rule, has to be opened to solve it. Photo courtesy of Pictura Paedegogica, The DIPF, "M. S. Christian Peschecks Arithmetical Solvency Key <Frontispiece>," Public Domain.

Son Remer's book was more elaborate and stepped the content up a notch. The introduction of son's manual addressed the many kinds of arithmetic that were taught and might well have corresponded to the three levels of arithmetic teaching dispensed at his school: "Some need arithmetic just for housekeeping, others for commerce, not only at home but also abroad, others still use it for scientific study, this is why I tried to organize this manual of arithmetic according to those different viewpoints" (Remer 1739: Vorrede). Remer the son indeed had more scientific aspirations and rearranged his father's material, embedding it into a more systematic account of arithmetic, including demonstrations. Whereas his father's contents were derived from the work of members of the *löbliche arithmetische Societät*, Remer son rather drew on some contemporary authors of encyclopedic and systematic expositions of mathematics such as Christian Wolff, Johann Michael Poetius or Christlieb von Clausberg. He used their works to reformat (and repurpose) the content of his father's book. It was indicative of a trend towards a more general and systematic teaching of reckoning and arithmetic, focusing first on properties of numbers and notations before embarking on the many specific and special rules.

When Carl Friedrich Gauss' grandfather, Jürgen, acquired citizenship of Brunswick in 1739, this entitled him (but also obliged him) to send his children to school. The son Gebhard Dietrich would thus visit one of the many schools where writing and arithmetic were taught, almost certainly the official *Schreib- und Rechenschule* near St. Catherine's, the same school Carl Friedrich Gauss would attend a generation later. Here, Gebhard Dietrich would learn enough arithmetic to be able to do the calculations and pay his journeymen or do some commercial arithmetic, an arithmetic thus tailored to specific needs. Indeed, Gebhard Dietrich would later help out a merchant, and in yet another anecdote, he is calculating the fees for his journeyman when, allegedly, 3-year-old Gauss squeaked out and remarked his father made an error in his calculation.

Gebhard Dietrich's going to school apparently also occasioned that the Gauss family acquired a copy of Christian Stephan Remer's manual, *Demonstrativische Anweisung zur Rechen=Kunst* (1737). As books were rather expensive, they had a long half-life and stayed in the family for a long time, being perused by more than one generation. The classic example would be the family bible, but other books would equally be a collective asset of the family (cf. Wardhaugh 2012). Anecdote has it, that the first book young Gauss perused (1785), was exactly Chr. St. Remer's manual which he would call "mein liebes Büchlein" (my dear book). In other words, Gauss' first acquaintance with arithmetic was not in school, but rather within the family circle.

As Gauss entered the Schreib- und Rechenschule in 1784, he first studied religion and writing, arithmetic came two years later in 1786. In the meanwhile, from 1785 onwards, he started going through the family's copy of Remer's manual. Büttner who had become headmaster in the *Schreib- und Rechenschule* near St. Catherine's, followed the tradition of the school. He had enlisted an assistant among his pupils, J. Bartels, and seemed to have continued the proven method of copying out examples to the pupils and let them solve it. Summing a sequence of successive integers was a stock example in the Remers' manuals, and Büttner may have well found his cue in their books to propose the example to his pupils. He may, of course, not have anticipated a young pupil to have already gone through the manual and to have found there the rule to solve the problem quickly (viz., sum the first and last number of the sequence and divide by two). Büttner acted upon Gauss' surprise solution and promptly ordered a new book from Hamburg (probably from the newly founded *Handelschule*, commercial school, see Chapter 2) while the assistant Bartels started studying more advanced mathematical books with the young Gauss.

This would be the beginning of Gauss' path towards becoming the famous mathematician. But not only his first steps were helped by favorable educational circumstances, his next school experiences would develop and further his capabilities systematically. When young Gauss changed to the Latin School of St. Catherine's some years later (1788), he arrived at school where, quite unusual for the time, a proper professor of mathematics had been appointed in 1773. Johann L. C. Hellwig taught a number of advanced mathematical topics (logarithms, algebra, indeterminate problems), some of them *privatissime*, in private courses. The school also boasted a small library, for which pupils had to pay a fee, even those who, like Gauss, had a stipend. The stipend system in Braunschweig was organized through circles of persons of rank in Braunschweig (among them masonic lodges). In a more or less systematic way, these circles incarnated new forms of patronage. Being given a stipend was only one aspect of this patronage, it also gave access to the notable circles of Braunschweig and their socio-political and intellectual resources, such as private libraries and collections. The systematic intertwining of local patronage and scholarship made it possible for the socially lower classes to ascend to better positions on the basis of merit. In Gauss' case, though a meeting with the Duke of Braunwscheig is featured in his anecdotes, it is rather August Wilhelm Zimmerman, professor at the Collegium Carolinum and *geheimer Etatsrat*, who was the connection between the court and the talented pupil. Zimmermann paid Gauss' fees and books from the state's resources and provided access to his library and circle of friends and colleagues.

FROM INDIVIDUAL TRAJECTORIES TO DEMOCRATIZATION OF NUMERACY

For both literacy and numeracy, the immediate starting point of the educational trajectory in the eighteenth century was the family circle and, by extension, the neighborhood and the family's relationships. Cultural skills such as reading, writing and reckoning, once acquired, were frequently transmitted to friends, family and neighbors. The elder reading to the younger, the mother or father teaching the children, or children imitating the local merchant, etc. This transmission often only imparted partial or imperfect skills, but accounted for a very basic dissemination of literacy and numeracy. In this transmission, the skill but also the context in which it was embedded, was communicated, too. For reading this often was the religious context of reading the Bible, for arithmetic, this was often a specific practice, with a custom-tailored form of arithmetic, such as a repartition rule (*Gesellschaftsrechnung*) or commercial arithmetic in the case of Gauss' father. A consequence of this transmission of skill through next of kith and kin is a long lasting continuity of practices. No school existed to become an arithmetic teacher (*Rechenmeister*), rather the profession was passed on from father to son, or from schoolmaster to assistant, reinforcing existing practices.

Increasingly, books, and later on periodicals, would play a part in this family circle. Books often remained for generations within the family, those able to read, reading out for those unable to read, those able to write, adding comments in the margins or on the blank pages of the book. In the eighteenth century (and later), the symbolic capital of a family is partially reflected in the books they possess. The family bible is probably the most common example of this, but also almanacks, dictionaries and manuals were equally handed on through the generations. The family circle thus became a locus of reproduction of this printed knowledge, as not only the copy was perused by reading, both silently for self-study and aloud for social entertainment, but also because comments and explanations were added and accumulated either on the pages of the book or through accompanying oral instruction. Books could also be passed along to friends and family, or read together.

While theology and politics were certainly the dominant discourses of the day, mathematics had its stable niche in eighteenth century literature. Commercial arithmetic, calendars, and also riddles and curious questions stemming from recreational mathematics were common topics. Rather than stimulating reflection and discussion, as did theology and politics, mathematical topics pushed for emulation of examples. This explains also why the so-called recreational mathematics remained an important genre in mathematics (see also Chapter 1 in Volume 3). The problems were set as entertaining challenges that dragged the reader in and incited them to participate in mathematics and display their prolixity. The attraction of the mathematical problem could lie in the story that was woven around the mathematical question, or it could lie in the challenge of the problem, asking for either continued and long computation, or for clever tricks and shortcuts, or even for the application of some algebra. The mathematical questions of England's popular almanac *Ladies' Diary* (1704–1841) and the answers the readers sent in were similar in this regard.

They all are witness to a participative form of reading, of engaging in conversation with the printed word.

The circulation of books (and later journals) went beyond the circle of family and acquaintances as new forms of sharing books were established during the eighteenth century. Some public spaces, such as coffee or tobacco houses but also parks and corners, became places where people gathered to listen to the latest printed news. Reading societies became established where, by paying a small fee, one could become a member of the society. This membership meant access to the society's library and a say in the society's book acquisition policy. Loan libraries were set up, schools created libraries, and private libraries flourished. This helped to make books (and periodicals) become ever more available during the eighteenth century. This growing circulation of printed books accounts for one of the major phenomena in education during this period: the autodidact (Bosse 2012). Whereas many educational situations were entrenched in long-standing habits and practices, the autodidact jumped out of this logic of continuity by reading the latest publications, by reading about other countries, by comparing books, etc. As a result of these reading experiences, the autodidact often came up with a new synthesis or with a reform proposal to overcome the generational inertia.

Not only for Gauss, but for quite a number of the best known mathematicians of the eighteenth century, the local network of friends and family combined with access to books constituted the backbone of their initial mathematical socialization. Leonhard Euler was introduced to mathematics by his father who had studied under Jacob Bernoulli. Euler deepened his knowledge autonomously by going through Michael Stifel's edition of Christoph Rudolff Coss (1553), the book inherited from his father. Later on, on the advice of Johann Bernoulli, he would study advanced mathematical books on his own, probably first by loaning copies from Basel acquaintances. Similarly, Alexis Clairaut learned mathematics under the direction of his father who had his own mathematical school in Paris. He later went on to study a number of important textbooks on algebra and the infinitesimal calculus (Nicolas Guisnée, Guillaume de l'Hôpital) on his own. To take another example, the blind mathematician Nicolas Saunderson first learned some mathematics from his father who was an exciseman, before continuing his mathematical education through the help of various friends who read out mathematical books for him. Even more of an autodidact was Johann Heinrich Lambert who after only an initial formation at a *Schreib-und Rechenschule* in Mulhouse, acquired the rest of his mathematical knowledge through books he would find through acquaintances. Later, when he earned a living as a private tutor in wealthy families, he gained access to their private libraries and perfected his self-education. In general, being a private tutor was a very common way for those with some intellectual training or capacity but without wealth to get through financially difficult periods. Such biographical paths are quite common throughout the eighteenth century and show the importance and variety of local networks of persons, learning, and books (see also Hans 1951). These include people from lower social classes (as Gauss and Lambert in the examples), or from classes already higher up on the social ladder. It also shows a shift from feudal patronage systems to supporting networks of citizens, both locally and within the republic of letters (e.g., through correspondence).

The example of Gauss' introduction to mathematics also illustrates the growing tendency of the states "meddling" in educational matters. In most regions, the schools were still headed by the local pastors and preachers, although the corner schools had been popping up in most cities. Many of the corner schools were run by people with some education who had not managed to get an appointment at some court, administration or church. Having failed to obtain a steady income in the form of a pension, or driven by the need to augment a paltry pension, they became free entrepreneurs in matters educational. To get by financially, many young men temporarily offered their services as private tutors for theology, language, mathematics and other courses. But next to the Church and free agents, the cities and states also developed an increasing activity in the educational market. This was, for example, the case for the Latin schools. The Church authorities (or, in the Catholic countries, particular orders such as the Jesuits or the Oratorians) organized many of the Latin schools (grammar school, *Gymnasium*, *collèges*, *lycées*), though often, especially in the cities, nobility, guilds, or charities had started creating schools, too.

In a generalization that sweeps the details and subtleties of the variety and regional differences away, one could say that in the bigger (say, over 10,000 inhabitants) and more internationally oriented cities, there were many schools and mostly operating under some government guidance, whereas in the countryside, especially in poorer, more rural and catholic parts, schools would be in poor condition and depend on the individuals working in and around the church. Similarly, state intervention in education, especially in trying to centralize and reform education on a grander if not yet national scale, can mainly be found in the bigger cities during the eighteenth century.

TOWARDS NUMERACY AS PART OF ALPHABETIZATION

The two events that accelerated this trend was the end of the Seven Years' War (1756–63) and the suppression of the Jesuit order in 1773. The Seven Years' War was both a local war, in Central Europe and around the Baltic Sea, and a global war, between the colonial powers of the day, especially between the Great Britain and France. In Central Europe, the end of the war and the ensuing financial crisis made many of the Central European states (Prussia, Saxony,

Austria-Hungary, etc.) realize they had to modernize both their administrations and their military. In France, the loss of territory to the Great Britain in North America and South India made clear they had to modernize their military, especially the marine. The year after the end of the war, France expelled the Jesuits (1764) and began a reform of its military schools, centralizing its administration and updating the (scientific) curriculum (see Chapter 2). For each class of schools (marine, engineers), a central examiner was appointed and the manuals were standardized to include more advanced mathematics such as algebra and infinitesimal calculus . These reforms and the gap left by the closing (or rebranding) of the Jesuit schools provided momentum for the French state to start centralizing its school system, but the virtual bankruptcy of the state prevented this from happening. However, it would have trickle-down effects on the schools that prepared for entering the military school and eventually prove to be the basis for the important reform of the Écoles normales and centrales during the French First Republic (1792–1804) after the French Revolution. In many Central European states, new military schools were founded after the Seven Years' War, most of them taking inspiration from the French model. More attention to mathematics and a selection based on merit rather than (or at least, next to) social descent were typical for these schools. They also had a trickle-down effect, it provided the stimulus for new (military) preparatory schools that, among other things, had more mathematics on the curriculum.

In the wake of the suppression of the Jesuits, most Central European states would double down on their efforts to modernize and centralize the school system, using the opportunity to put into place a central control organ and a common curriculum or examination on a national scale. The effect was not limited to the catholic states only. Even if states like Protestant Prussia or Orthodox Russia did not suppress their Jesuit schools, they still joined other countries in their desire to modernize, centralize, and control education. Many states would issue a number of laws in the later third of the eighteenth century, making attendance at elementary schools obligatory. They detailed both the content and the form of examination within the educational system and put reading, writing, and reckoning on the agenda. Especially the laws issued by the Emperor Joseph II in Austria (1784) provided a blueprint for the nationalization of the schools, both elementary and secondary. They worked towards a secularization of schools, stipulating among other things the usage of the vernacular instead of Latin. The central idea was that schools were a means to educate people into useful citizens for the state. In practice, the important result of this new wave of laws was that actual control instances were put into place that could enforce the realization of a common curriculum and of compulsory alphabetization, though it would still take long into the nineteenth century before that would be realized nationwide.

The institutional reforms and the education laws led to the publication of a great number of pedagogic writings, mostly manuals but also pamphlets, as well as to some experimental school forms. One of the early pioneering institutes was Johann Bernhard Basedow's Philantropic Institute at Dessau in the 1770s that advocated education based on realia (see also Chapter 7 in Volume 5). An essential novelty introduced there was the splitting of the manual in two parts, a part for the pupil and a part for the teacher. Instead of a book that could be used both by a teacher and by a (possibly autodidact) pupil, the teacher now had access to background material for his teaching, both methodologically and practically, that remained hidden to the pupil. This innovation happened in parallel with the (slow) establishment of teacher's training seminars. A proper teacher's training and its reflection in the pedagogic material marked the beginning of professionalization of education and the end of the free entrepreneurs and corner schools.

This is also attested in Johann Ignaz Felbiger's *Methodenbuch* (1777) written for Austria's Maria-Theresia and preparing the ground for Joseph II's reforms. Felbiger meticulously listed in the *Methodenbuch* how the teacher should proceed and the pupil should follow. Reckoning was to be taught after the pupil knew how to read and write, but, contrary to the corner schools where each pupil was working alone on their proper example, the calculating examples should now be done before the class by the teacher, and then repeated individually by the pupils on their *Rechentafel* (slate). This was to be repeated until all pupils could replicate the example.

In France, shortly after the Revolution, the Comité d'instruction publique (Committee for Public Education) had opened a competition to write elementary books for use in the elementary and central schools (1794). Nicolas marquis de Condorcet who had been preparing the educational reform (before being guillotined) started to write such a manual that would eventually, in J.-B. Sarret's version (1799), win that competition. This manual had two parts, one for the student and one for the teacher, and was quite programmatic. It can stand for many others in its attempt to repurpose classic arithmetic treatises and achieve something more general. Condorcet presented the elements of arithmetics as being at the same time the elements of logic, or rather, the laws of thought. This idea went back to the early eighteenth century (as for instance voiced by Locke), but Condorcet now made it more concrete:

> The elements of [arithmetic and geometry] should not only enable the children to execute their calculations surely and easily when they need them, but should also serve them as the elements of logic and help them to develop their abilities to analyze their ideas, to suspend or fixate their judgments, to reason correctly. (Sarret 1798: xxxij)

Applied to the teaching of elementary arithmetic, the creation of numbers by adding one became the mold of a logical proposition:

> [The teacher] will teach them that, to express a consequence, one uses the word THUS, and one calls REASONING the reunion of one or more propositions, seen as true, together with the proposition that is concluded from them. Let's take as example the number SIX:
>
> > SIX is formed by the number FIVE, who precedes it immediately on to which has been added ONE; SIX is thus greater than FIVE, because it is formed by FIVE and because one has to add ONE to FIVE to get SIX. (Sarret 1798: 5–6)

By showing the general reasoning principles behind arithmetic, one could at the same time introduce the pupils to logical thinking and avoid the pitfall of mechanically applying rules and formulas.

This double message, emphasis on the (psycho)logical structure behind arithmetic and the avoidance of mechanical rule-application, ran like a mantra through the pedagogic literature of the day. It achieved a number of ruptures with the tradition of reckoning manuals of earlier centuries (cf. Bullynck 2008). First, it dissolved the proximity of reckoning with writing, because they now started from the concepts and, often, with mental arithmetic before introducing the "numeration," the way of writing ciphers and operating with them. Second, it foregrounded general principles, thrusting the practical arithmetic aside. As a consequence, the many special rules disappeared, and the four arithmetic operations (and mostly the rule of three) were the main focus. Finally, elementary arithmetic, detached from its practical context, slowly achieved a propaedeutic value, it became indeed akin to logic.

One of the most extreme but also most influential pedagogues to implement this turn was Johann Heinrich Pestalozzi. Together with his collaborator Hermann Krüsi, Pestalozzi developed a systematic approach for teaching elementary arithmetic in their experimental school of Burgdorf (1799–1803) and later in Yverdon, both in Switzerland. The cornerstone of the method was its development of unity. As Pestalozzi wrote in his programmatic novel on education *Wie Gertrud ihre Kinder lehrt* (*How Gertrud teaches her children*:

> Arithmetic wholly springs from the simple composition and separation of multiple unities. Its elementary form is this: One and one is two, and one from two stays one. Every number, whatever its name, is nothing else but an abbreviation of this essential *Urform* of counting. (Pestalozzi 1801: 259)

All numbers and all operations spring from unity and go back to it. The numbers and their transformations were exercised using the table of unities, taking two unity strokes makes for two, and two strokes make for two unities.

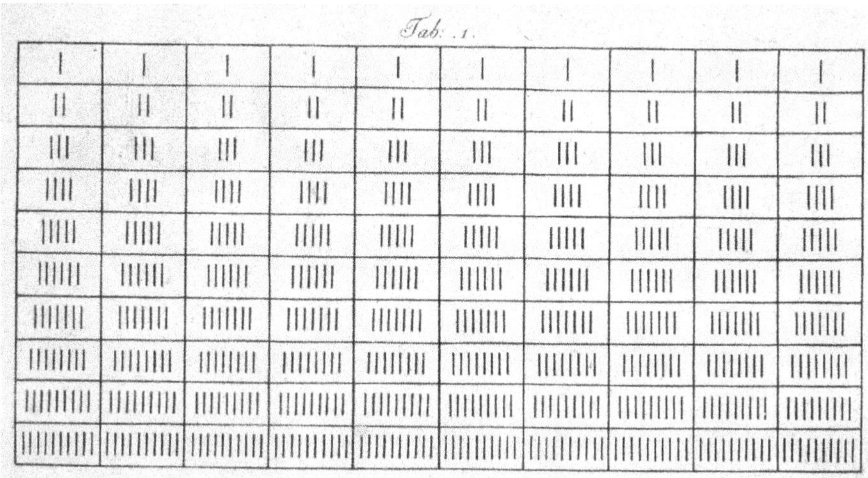

FIGURE 1.4. J. H. Pestalozzi's Table of Unities. Photo courtesy of Pictura Paedegogica, The DIPF, "The table of units (1803)." Public Domain.

In painstaking detail, according to the principle that the psychology of the child advances through smallest differences, Krüsi and Pestalozzi published a multi-volume handbook with mantra-like series of numbers and their transformations. It started from expressions like, "One from three is one from two and one, is one from one and one and one, is one and one, is two." But it managed, in the later parts on "parts of numbers" (fractions), to solve question such as:

> 48 times 1 is 4 times the 7th part of how many times 9?

with an answer defying common understanding:

> 48 times 1 is 4 times 12, 4 times 12 is 4 times the 7th part of 7 times 12, 7 times 12 is 84 times 1. 84 times 1 is 9 times 9 and 3 times the 9th part of 9

the answer thus being 9 and a third. While these mantras were often ridiculed by contemporaries, they consequently showed how the number one is the pivot around which the method developed the concept of number. The handbook was for the teacher, the numbers and their transformations had to be recited by the pupils together and then exercised using the table of unities. This method completely disconnected arithmetic from writing. Learning to write down numbers and compute with them happened only much later in their curriculum, only after the concept of number was firmly understood.

FIGURE 1.5. A caricature of Pestalozzi's method for teaching arithmetic. The pupils hold the table of unities in their hands, one pupil despairs and pulls out his hair. The inscription beneath is a travesty of Pestalozzi's method and reads "10 times 10 is 10 times the 9th part of 9 times 10." Photo courtesy of Maarten Bullynck.

Though Pestalozzi's method would not be copied literally, it would have much impact on later handbooks, especially for the early stages of teaching arithmetic. Mental arithmetic would precede reckoning, and apprehending the concept of number would precede any concrete application.

Pestalozzi's extreme form was in part dictated by his desire to make teaching accessible to non-teachers, in particular to mothers, often unschooled themselves, that wanted to educate their children:

> a schoolbook [. . .] must be arranged thus that the unschooled man and even the mother can, following its thread, have enough to rely on, so that they will always lead the child in its progressive development, so that they will always be one step ahead of the child itself. (Pestalozzi 1801: 64–5)

As Pestalozzi's ideas were tried and slowly integrated in practice, and for use by actual teachers, only the initial stages of his method would remain. This would be done by a number of pedagogues in the German states: Ernst Tillich in Saxony, Joseph Schmid in Yverdon, Wilhelm von Türk in Prussia, or Heinrich Stephani in Bavaria (see also Chapter 1 in Volume 5 for the uptake of Pestalozzi in the Anglo-Saxon world). While they would dump the mantras, they would keep the possibility of composing and decomposing a number in a multitude of ways, because this offered a trapdoor for reintroducing the classical tricks and shortcuts in reckoning. Thus, 30 might become 10+10+10, or 9+10+11, or 1+ 9 1/3 + 9 1/3 + 9 1/3, depending on the kind of reckoning advantage that could be exploited. But instead of rote learning all special tricks and cases, the pupils were now trained to split up a number in a variety of ways. This splitting up was considered a general capability of the mind, a competence that could be applied later on when confronted with a specific case.

PRACTICAL ARITHMETIC AND PRACTICAL GEOMETRY

Before the introduction of a more general form of arithmetic as part of compulsory alphabetization, eighteenth century numeracy, in continuity with previous centuries, was often of the practical, concrete kind. Manuals of practical arithmetic first gave the definition and operations of decimal numbers, that is: numeration (the writing of numbers in the decimal positional system), then the four arithmetic operations, (addition, subtraction, multiplication, and division), and finally, the rule of three or the golden rule. These books continued with the same operations, but now explained for fractions, before getting into the many special rules of trade and business involving concrete numbers with money, weights, and measures (see also Chapter 1 in Volume 3). Bookkeeping was often added to the practical part. Nearly for each professional practice, a specific kind of numeracy existed. Each kind of practical numeracy revolved around the concrete problems of the trade and profession and brought in the tools and tricks needed in the field or on the workfloor. It drew on a selection of useful devices and rules from mathematics, that were subsequently specified, detailed and subdivided for application to those problems.

To get an insight into how teaching, commercial arithmetic, local laws and measures and surveying were intrinsically tied up, the journals of a local school teacher between 1771 and 1792 might be telling (Bernet 2000). Delahaye was the local clerk and teacher in Silly, a small village in Ile-de-France of some 580 inhabitants. He did not only teach, but was in general the go-to man for all things writing and reckoning. In particular, he was a part of the parish council (that also functioned as village council), functioned as the local clerk, in charge of their accounts and of surveying. For the farmers, Delahaye made a "mesurage figuré" (measured map) measuring up the land "divided in 80 pieces" in *arpents*, *perche*, etc. (with an *arpent* being 100 *perches* and a *perche* being 22 *pouces* according to the local measuring). Although Delahaye did the surveying, it was the school teacher and *arpenteur juré* (sworn surveyor) of the bigger village of Plessis who had to sign the plan. Delahaye was paid 60 *livres* in two installments for making this plan (with the *livre* being *60 sols* and a *sol* being *12 deniers*). He further collected the *dixme*, the tax due to the local priest, for instance, 4 *sous* for each lamb, with 237 lambs in the hands of six villagers, Delahaye got 6 *livres* for his work. He also drew up an act to divide the heritage of the dead Jean Charles Vigneron, again, he got 3 *livres* for his work. In the meanwhile he bought or received *queues* (=*432 pintes*) of wine, *setiers* (=6 *minots*) of grain, or *cordes* of wood, and subsequently sold parts of it or gave it away to compensate for goods and services he had received from other villagers. The list could be continued endlessly, but the point is that Delahaye was not only a local teacher, he was the local professional for all things mathematical, be it commercial arithmetic, accounting, surveying, etc. In all his actions, the peculiar aspects of the local economy and exchanges surfaced. As a consequence, his journals testified of the transformation of land into taxable entities, of taxes into goods, of money into goods, of goods into services, of goods into other goods (possibly from neighboring villages and regions), of services (in particular his clerical and educational services) into goods and money again. The kind of arithmetic and geometry needed was not just inscribed in this network of local relationships, it was underwritten and conditioned by it. This was, for many places during the eighteenth century, (still) the reality of numeracy. In this sense, each locality generated its own peculiar numeracy.

The material support for doing mathematics could be quite different for each practice and profession. Farmers often stuck to cutting notches in wood, local merchants remained attached to calculating on abacus-like lines on a table, while astronomers were conversant with decimal fractions and logarithm tables as daily work instruments. Similarly, the plainest of surveyors had his surveyor's chain, the better trained surveyor used a plane table, and the astronomer used his zenith-sector equipped with a system of lenses and micrometers. There are a number of trade-offs involved, the more complex the instrument, the more skill and patience needed to achieve more precision, and the more mathematics is needed to use and reduce the data obtained (see also the example of Picard's leveling instrument in the introduction). For everyday use, robustness prevailed, with less precision and certainly less mathematics.

There was a close but subtle correlation between the amount of mathematics brought in, and the details of putting that knowledge to work in practice. The more a practice was tightly defined and minutely described, the less mathematical generality was needed. Also, the mathematics involved might be implicit, viz. materialized or even incorporated in the form of the mathematical tool. A notable example of this is the implicit use of a decimal system in the practice of surveying. Already in 1585, Simon Stevin had proposed in his *Disme* to incorporate decimal fractions in the measuring instruments of various trades, viz. the measurement of tapestry, of wine (gauging) or of lands (surveying).

Though decimal fractions were still absent from many trades in the eighteenth century, in many countries it had been implicitly introduced in surveying through its instruments. The surveying chain in the Netherlands, in the German states or in Great Britain all had been decimally subdivided. In Great Britain, for instance, it was known as Gunther's chain and subdivided in 10 times 10 links, one chain corresponding to 4 poles or perches, or also, 22 yards or 66 feet. All measurements with the chain were thus decimal and duly noted in the field-book. Only after all the measurements and calculations were done in the field-book, would the results be converted back to local measures for the official document the surveyor had to deliver. The notation in the field-book could have many forms, as transpires in the following quote from John Love's *Geodaesia*:

> When you put down in your Field-Book the length of any Line, you may set it thus, if you please, with a Stop between the Chains and Links, as 15 Chains 15 Links 15.15. or without, as thus 1515, it will be all one in the casting up. (Love 1688: 56)

Love's manual was popular in the Americas, first printed in 1688 but with numerous reprints until the end of the eighteenth century. The quote brings out that lengths may be noted either materially as parts of the chain (15 Chains 15 Links), concretely as expressed in measures, as an integer number (1515) or as a decimal fraction (15.15). The computations with the chain measurements were likewise open for interpretation as a calculation with concrete numbers, with integers or with decimal fractions:

> Let A be a true Square, each side being 10 Chains; multiply 10 Chains 00 Links by 10 Chains 00 Links, facit 1000000. from which I cut off the five last Figures, and there remains just 10 Acres for the Square A. (Love 1688: 122)

This ambiguity was intentional, as it it had to address a rather mixed audience of surveyors, with different educations and thus each with a different set of skills and notions. The synchronous plurality of practices was thus present in the manual itself (not unlike the manual of Remer son that had different audiences in mind).

Love's book also illustrates the trade-off happening in practice between mathematical knowledge and mathematical tools. Indeed, for "those that do not well understand Decimal Arithmetick," Love suggested that they would not try their hands at performing a division (by 66, for converting the chains into feet), but would rather use "a table" inserted in the manual where they could seek the number and write down its quotient. Similary, for computing a square root (useful in Pythagoras' theorem), Love gave a simple arithmetic algorithm, but also mentioned it could be done "by the Logarithms or Artificial Numbers, [it] is very easie and pleasant, but Surveyors have not always Books of Logarithms about them." Further, in case you would need to find a third side with a given angle and two given sides, Love explained one could find it out "by laying them [the sides and the angle] down thus upon Paper," and then physically measure the third side on paper (scaled to the right proportions). One could also use "Gunter's Scale," a special-purpose sector for the surveyor, or one could use the trigonometric tables included at the end of the manual, together with the logarithms (Love 1688: 106). The differentiation of methods corresponded to the kind of surveyors. Where mathematical knowledge was lacking, a trick or device existed that could supplant the missing knowledge and still solve the problem at hand.

Numerical tables were very prominent aids in this regard, they were part and parcel of practical arithmetic. They gave numerical values that could be retrieved without calculation and without the mathematical background needed to calculate them (Wess 2012). Another form of the same idea, were the sectors (*Proportionalzirkel, compas de proportion*), a mathematical instrument that combined two rulers joined by a hinge, each ruler having one or more graduated scales. It could be used to translate a scale into another scale using the property that similar triangles have proportional sides. The scales could be logarithmic, trigonometric, perspectivistic etc., but also practical scales, mapping numbers to their corresponding values useful in gunnery, excise, surveying, navigation, etc. Sectors and tables both alleviated the mathematical needs of the practitioner. They substituted a (complex) calculation for a simple lookup or a simple geometrical operation. The mathematical knowledge was embedded in the instrument, device or table and made accessible to direct manipulation without further ado.

Another example from another field, the military, illustrates once more the correlation between the level of numeracy (or mathematics) of a profession and the tools applied in the field. The obvious interest of the royal patrons in obtaining more accurate firing tables had motivated the academies of Berlin, Paris, or at Woolwich to conduct much research on the integration of the ballistic equations and on the measurement of the empirical values that had to be put in these equations. Technological advancement (guns with higher velocity), new experimental techniques (Robbins' ballistic pendulum, cf. Chapter 6), and the power of the new infinitesimal calculus now enabled for a more complex, but more precise calculation of firing tables. A side effect of this professionalization in ballistics was not only that the tables were difficult to compute, but more even, that they became also difficult to use, as they were no longer simple one-entry tables, but had become a set of tables (or a double-entry table if combined in one). The use of these tables and the computations needed in using them (interpolation and determining the species of the trajectory) very nearly exceeded the mathematical abilities of the common artillery officer. To solve this practical problem, J. H. Lambert in Berlin came up with the idea to translate the tables into a visual format. He took inspiration from the sector and published his "échelle balistique" in 1765. It is one of the first examples of what would later be known as nomographs, graphical tables where numerical values could be read by following the curve (see also Chapter 6 in Volume 5 and Tournès 2022). The underlying mathematics got exchanged for an intuitive, but implicit materialization of that mathematics.

More intricate arithmetical instruments, however, such as those for doing addition, multiplication or differences through a combination of cog wheels and gears, did not enter everyday practice. They were too costly and technologically too difficult to be produced *en masse*. One of the technological advances in this field was the so-called Leibniz wheel or stepped drum, used in Leibniz's calculator of 1671 that could perform the four species. The Leibniz wheel had teeth of different lengths and could thus steer and control not one cog wheel, but a set of different cog wheels. The Leibniz wheel led to the development of more intricate calculators, for computing differences and large tables. But these calculators were rarely used in practice, but rather were displayed in the curiosities cabinets of the rich and noble, or in the public spectacle of science in the salons and in the universities.

As practices changed, due to technological innovation, scaling up or growing internationalization, the amount (and nature of) the mathematics needed would change, too. Such shifts could be witnessed in a number of contexts during the eighteenth century, for instance in surveying or ballistics as discussed before, but also in commercial arithmetic.

Commercial arithmetic goes back mostly to Italian Renaissance sources and did not change significantly content-wise in later centuries (cf. Chapter 1 in Volume 3). It contained the rules of arithmetic with integers and with fractions, though not with decimal fractions. It further enumerated many calculating rules for the specific situations of commercial arithmetic: computation of tare weight, discount, barter, loss and profit, mixtures, exchange rates, interest and annuities. With the mathematics of commerce came a lot of specialized vocabulary of commerce. For instance, for the popular arithmetic of exchange bills, one needed to know what a remittent, a trassent, a praesentant, and an acceptant were. The remittent gave cash money to the trassent who returned and issued a bill of exchange to the remittent, the praesentant then presented the bill to an acceptant who accepted to exchange the bill for cash money. As each of these transactions could involve exchange from one currency to another, and commissions for these exchange, the arithmetic could get involved. The gradual complexification of financial products during the eighteenth century made it necessary to combine rules or introduce longer, more intricate rules. Compounded interest and annuities demanded long calculations (cf. Chapter 4). Also the chain of compounded money exchange rates in a internationalizing world of trade stretched the traditional commercial mathematics to its limit. The following example, quoted by Abraham Gotthelf Kästner in his discussion of commercial arithmetic, makes this clear:

> A *pipe* of olive oil from Cadiz is 84 *pesos de plata*, the costs are 1 *peso* per *pipe*: the broker's commission are 2 ½ percent, insurance is 5 percent. The exchange rate from Cadiz to Amsterdam is 98 *Pfennige* banco for 1 *Ducado de Cambia* of 375 *Maravedis de Plata* [one maravedo=272 pesos], and the exchange from Leipzig to Amsterdam is 141 *Reichsthaler Louisd'or* per 100 *Reichsthaler* banco. The costs in Hamburg are 1 ½ *Reichsthaler*, in Magdeburg 1 *Thaler* 8 *groschen* and in Leipzig 19 *groschen* per 1 netto *center* [with a *pipe*=832/110 *center*]. (Kästener 1786: 314)

In the original question (1783), a special rule was posited to find the answer: add the cost of 1 peso, then multiply the exchange rates with each other and than divide by 97003. In Kästner's discussion, however, all exchange rates are duly translated as proportions and reduced to decimal fractions, the unknowns given as algebraic letters. The actual calculation then is done through logarithms. Indeed, the writers and compilers of manuals of commercial arithmetic had been prompted to include either more advanced mathematics or more extensive tables in their works. Logarithms (and decimal fractions) started to appear in manuals for commercial arithmetic from the early eighteenth century onwards to abbreviate the many multiplications and divisions needed. Also some algebra was introduced as a short and concise way of reasoning with many numbers and quantities. More often still, precalculated tables of interests, annuities, exchange rates, etc. were printed to relieve the merchant of the many calculations and reduce the mathematics to looking up values.

INTRODUCING THE METRIC SYSTEM

Another big shift in practical numeracy was prepared in the eighteenth century, one towards a general and international system of weights and measures. As described before, measures, weights, and currencies varied from region to region, from village to village even, and much of the teaching of practical arithmetic and geometry in small local schools went into profusely describing the local customs and how to apply specific mathematical procedures to those. The extra courses were then often reserved for opening up practical arithmetic and geometry to the usage of the measures, weights and currencies of surrounding regions, albeit against payment of an extra tutoring fee. But even there, the teaching of more general rules and notions of mathematics was hampered by the unruly detail of the many local habits and customs materialized into idiosyncratic systems of weights, measures, and currencies.

Throughout the eighteenth century, it had been a common topic of Enlightenment discourse that such remnants of feudalism should disappear and make place for a rational and general discourse, but this discourse had largely remained disconnected from local practice. The salient point was that the issue was intricately interwoven with the legal and political balance that was kept between the centralized administrations and the local authorities. Even in the late Middle Ages, establishing a royal monopoly on the minting of coins had only been achieved by buying off the local *seigneurs*. Therefore, attempts at introducing new weights and measures from above, by a central administration, had few chances of working out. At best, they would add to the chaos by introducing yet one more measure or weight that would only have a legal existence but no embedding in any practice.

Nevertheless, the idea of a universal system of weights and measures slowly gained acceptance in another community, that of the scientists (see the Introduction). In the Paris academy, they proposed in the late eighteenth century a project of establishing a universal system of weight and measures based upon the metre (defined as the 10-millionth part of a meridian going from the poles to the equator) and built around the powers of ten, just as the decimal positional notation for numbers. The First French Republic (1792–1804) adopted the project and made its use and teaching obligatory to meet its own ambition: to get rid of the weights and measures of the Ancien Régime. The new system was extolled for its "democratic" virtues and could potentially open up the teaching of elementary arithmetic and geometry beyond the simplest additions of integers. As one of the most prolific and influential mathematical textbook writers of the early nineteenth century, Sylvestre Francois Lacroix, pointed out in the lemma *Mesures* in the *Dictionnaire raisonné et universel d'agriculture* (Rational and Universal Dictionary of Agriculture, 1809):

> The difficulty of these operations [computing with the old measures and with ordinary fractions], nearly unknown in the local elementary schools, accounts for the fact that the overwhelming majority of people who can read and write know no other arithmetic rules than addition and subtraction. [. . .] If decimal computation could be introduced in elementary schools together with the usage of the new measures [of the metric system], than not only the housewife would be able to do all the calculations she needs, but also the worker could do all his measurements without pain; adding the use of ruler and compass to trace some geometric figures, he would be able to design his own plans, and the farmer would find no difficulty in surveying his fields. (Rozier 1809: vol. 8, 293)

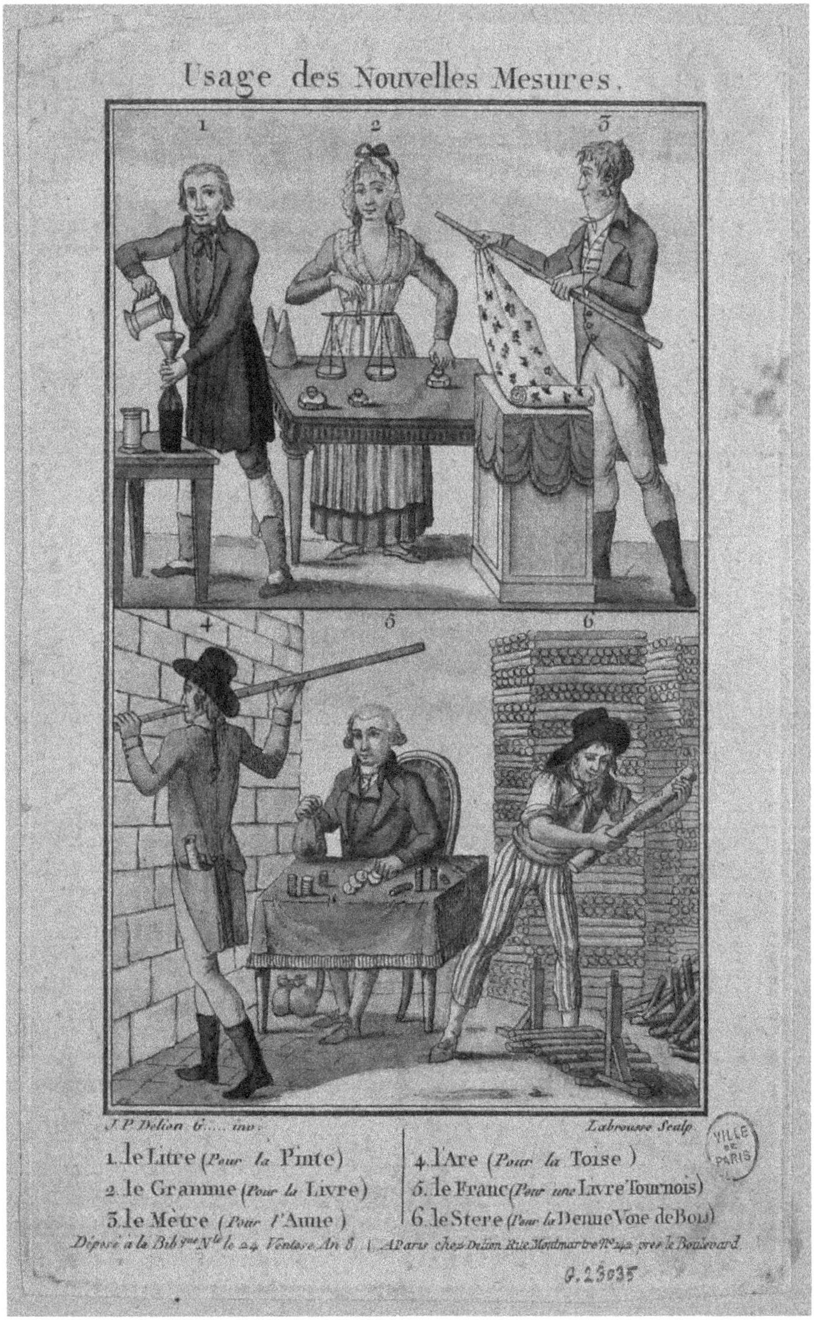

FIGURE 1.6. A poster commissioned by the Comité National to make the new metric system known to the public. The names of the new measures are given, the names of the old measures are in parentheses. Photo courtesy of Musée Carnavalet, Histoire de Paris, Public Domain.

But while the new system was deemed to emancipate every person in the republic from the feudal shackles of the old weights, measures and laws, the practitioners and the people complained that the new system shook up their habits. The writer Louis-Sébastien Mercier satirized the new system, saying that the people did not need a meridian, let alone an academy, to use the old measures. Meeting such resistance among the population, the metric system was even abandoned in France for a while. It would take some decades well into the nineteenth century before the metric system would be generally adopted in Continental Europe.

CHAPTER TWO

Practice and Profession

Embedding Mathematical Practices in the Modernization of the State

MAARTEN BULLYNCK

While in many ways the long eighteenth century can be seen as a continuation of the centuries before, there are quite some changes happening in the social fabric and in the communication patterns that underwrite and support mathematical practices and professions. The accumulation and confluence of those changes will slowly prepare the way for the nineteenth-century institutionalization and profesionalization of mathematics featuring the new social phenomenon of the professional mathematician (see Chapter 2 in Volume 5). In previous centuries, it is more accurate to speak of mathematical practitioners or of trades and professions that make use of (some) mathematical knowledge and skill (see also Chapter 2 in Volume 3). This chapter will visit a selection of places where mathematics were used, from the lofty academies over places of higher education to trades and practices using mathematics on an everyday basis. Special attention will be paid to moments where changes in local practices occurred, for instance, the time before and after Austrian Succession War (1740–8) and the Seven Years' War (1756–63). This will allow us to describe mathematical practice in motion while embedding them in their immediately practical, economic and political contexts.

ADMINISTRATION, COMMERCE AND INDUSTRY

Economy and Administration

Due to the many wars of the eighteenth century, many states were in a permanent state of near bankruptcy and were always looking for ways to increase their income. New taxes were raised to finance the wars, such as the excise tax in England (cf. Chapter 6), or the tax on playing cards to fund the École militaire in France. Many states also tried to reform the basis for one of the principal incomes: land tax. In the Netherlands, new annuity schemes were set up to finance the military defense. The establishment of state lotteries was also a popular way to finance things. Finally, states sought to further the exploitation of the natural resources in agriculture, forestry, and mining, and fostered industrial development and the commercial expansion of the trade routes. It must be noted that in most cases, states did not directly sell and negotiate the trade, industry or commerce monopolies, but sold them to a kind of private-public companies. These advanced the money the states needed and in return they could exploit, sell and also enforce the monopoly. Colonial trade companies were one example, la Ferme Générale which collected the taxes in France another. Most factories, manufactures and mines were equally held by both private investors and public servants or noblemen.

On top of that, states were increasingly woven into a global economy. This was especially true for the colonial powers Spain, Portugal, France, Great Britain, and the Netherlands, but also for the imperial powers of Russia, Austria, or Prussia and the many smaller states that were, through alliances and wars, linked up with a globalizing market. The main financial center for banking and insurances in the eighteenth century was Amsterdam, followed by London. The many wars and the challenges of a global market made for many financial bubbles and crashes in the eighteenth century. Most famous are South Sea Bubble and the Mississipi Bubble, both 1720. These bubbles were the results of schemes to finance the debts of Great Britain and France respectively by granting trade monopolies for the colonies to the companies that advanced the money to pay the state's debts and budget. The Amsterdam banking crisis of 1763 then was a crash of the financial market after the Seven Years' War, causing financial crises in Hamburg and Berlin, and announcing the financial collapse of the Dutch state near the end of the eighteenth century. As a consequence, the center of financial trading would shift from Amsterdam to London in the nineteenth century.

In such an economic context, it need not wonder that new economic theories were worked out, abandoning the old mercantilist idea of accumulating money reserves through positive trade balances. Since the late seventeenth century, there had been efforts to systematically collect and use numerical and statistical data to help the state assess its economics. These data would then be interpreted with the aid of the legal, philological and mathematical sciences. In Great Britain, Cromwell had commissioned William Petty to make a new cadaster for Ireland, which led Petty eventually to propose a political arithmetic, using demographic data to improve public accountability. In France since 1670, Colbert had also ordered that a systematic account of demographic evolution be held. In this vein, Sébastien Le Prestre de Vauban, famous for his fortification engineering, had proposed to reform France's complicated tax system and replace it by a *Dixme Royal* (1707), the king imposing a tax of a tenth on all fruits of the land and on all revenues. In his proposal, Vauban detailed the forms to be used for the census of the population and for systematically enumerating all resources for each parish.

Vauban's plan was not taken up, but two generations later, the French so-called Physiocrats would also take (agricultural) labor as the basis of value instead of precious metals and would propose a reform of the economic system. They garnered political influence after the financially disastrous Seven Years' War. A generation of *administrateurs-géomètres* (administrators-mathematicians), including François Quesnay and Anne-Robert Jacques Turgot, would try to alter the administration and its planning, thus modernizing parts of the French taxation system on a basis of data and calculations. Turgot's disciple Nicolas marquis de Condorcet would later lobby for establishing a proper *Bureau d'arithmétique politique* (Office for Political Arithmetic, 1791) under the First Republic.

This evolution was intrinsically linked up with the development of the financial market and its products. As the complexity of financial products increased, the bankers were among the first to enlist mathematical expertise to analyze them. The states, belatedly, had to follow to keep up with the bankers. The point became most political when Jacques Necker, a Geneva-born banker who had become in charge of France's finances between 1776 and 1781, took up debts in Geneva's banks through life annuities. The bankers were well informed about the latest mathematical models of life expectation and the effect of inoculation. They had developed new, intricate life annuity products, for instance, one on the heads of 30 inoculated little girls. When Euler wrote 1778 to Turgot, Necker's rival in matters of France's finances, on his newest work on life annuities, Turgot complained it came too late. Indeed, Turgot was no longer in charge politically, as Necker had already took up credit in Geneva through the newest annuity schemes, but more even, Euler's mathematical modeling could not hold up with the rather twisted particularities created by the Geneva bankers' inventiveness.

In the German states Cameralism (*Kameralwissenschaften*) proposed a modernization of the chambers that governed the state, calling for centralization and statistically informed decisions. Authors in the Anglo-Saxon world rather opposed the French or German tendency towards centralization and involvement of the state, but shared the interest in collecting and analyzing data. Instead, free trade and a minimal role of the state were central, as voiced most influentially by Adam Smith in his *An Inquiry into the Nature and Causes of the Wealth of Nations* (1776).

The slow transformation of the chambers into a more modern form of central administration, including the re-nationalization of outsourced activities such as monopolies or taxation (e.g., France's *Ferme Générale*), only happened in the nineteenth century. But throughout the eighteenth century, the chambers tried to compile more accurate statistics, especially census data, and started surveying the land to update the land registration system (*cadastre*). Once more, the short-lived First Republic (1792–1804) showed the ways of things to come, founding institutes to compile statistical data, to modernize the land register and opening up the administration to trained persons of merit rather than to those with the right genealogy and connections. These activities did not require very advanced mathematics, but the systematicity needed and the sheer amount of data involved prompted reflection on how to deal with errors in the data and how to check the consistency of the data. In part, this was also a discussion about the place of mathematical method in compiling and analyzing these statistics. Whereas some would stick to the older literary or philological approach to compiling statistics, generating descriptions and summaries of facts, others, especially those mathematically trained, would advocate for computations, calculating means and using early statistical sampling methods (Desrosières 1998: 35–40).

Next to the *Bureau de politique arithmétique*, founded in 1791 at the insistence of Condorcet and Antoine Lavoisier, a *Bureau du Cadastre* was created in 1791. Under the direction of Gaspard Riche de Prony, the institute was organized to comprise mathematically trained persons (coming from the École polytechnique or the École des ingénieurs-géographes) and low-level calculators. In the absence of new surveying data, de Prony decided to embark upon another project to keep the Bureau busy: the calculation of new logarithmic and trigonometric tables. From 1793 to 1797, the Bureau worked on those tables, employing 8 mathematicians and 14 calculators, the latter approximately doing 1,000 additions a day per person, whereas the former prepared the calculating sheets and checked the obtained results. A group of mathematicians "of high merit," including de Prony and Legendre, had prepared the mathematical formulas and the plan of computation. According to de Prony, this division of labor was inspired by Adam Smith, though similar organizations of computing labor had been used by astronomers such as Joseph Jérôme Lalande long before. His project, however, would later inspire Charles Babbage's mechanization of mathematical table making in the form of the Difference Engine (1821) (see also Chapters 4 and 5 in Volume 5).

Industry, Technology, and Business

Early industrialization also began in the eighteenth century, advances in milling, textile production, mining, or the use of steam power. There was a general scientific interest in all sorts of machines and in manufacture or industrial processes, as witnessed by their presence in encyclopedic works or by the discussions in the academies. However, the invention and development of these technologies and production processes were largely not driven by or based on systematic scientific or mathematical method. If science or mathematics were invoked, it was often more for promotional purposes, e.g., by *Projektemacher* (project makers) who wanted to convince the courts to pay them a pension or to finance often unfeasible plans. Nevertheless, informal ways of communication between persons, and also between theory and practice did exist. Ranging from contacts between universities and industrial sites, over the touring of industrial sites to the establishment of informal societies. The most famous of these societies was the Lunar Society of Birmingham (1755) where inventors, scientists and others informally exchanged each full moon, but many other societies and clubs locally provided places of encounter for those who were interested in science, experiment, and applications.

The more classical sites of learning, academies, and universities, mostly did not engage in actual application of science though this slowly changed in the latter third of the eighteenth century. Only some universities in the periphery or those close to industrial sites such as mining sites did offer courses on applied sciences, e.g., chemistry, mining, or forest science. Important examples are the Scottish universities Glasgow and Edinburgh, the Swedish university of Uppsala, or some of the German universities. After the Seven Years' War, a number of special schools were founded. Mining academies were opened in Freiberg (1765), Schemnitz, Berlin (1770), and Clausthal (1775). A cameral school was started up in Kaiserslautern (1774) but was later integrated in the university of Heidelberg (1784). Forest schools, applying cameralism and mathematics to the exploitation and conservation of the forests, appeared in Wernigerode (1763, later moved to Ilsenburg), Zillbach (1785, later moved to Tharandt), in Munich (1780) and in Schemnitz (1808). These schools all tried to systematize the education of the persons involved in mining, forestry or general administration of the state. Instead of the transmission of know-how through apprenticeships, or the simple appointment on the basis of noble descent and some general university study, the new schools provided a systematic curriculum combining theory with practice, including quite some mathematics. Especially practical geometry, a topic also taught at most military schools, was useful, with application to the real-life situation in mining or forestry. But also arithmetic (including decimal fractions) and even some algebra was part of the training. Trigonometric functions or analysis, however, mostly stayed out of the picture.

Another new school form that was born during the eighteenth century was the business school. Those who wanted to pursue a career in commerce and business normally attended a writing and arithmetic school, either a corner school or a city school. Sometimes they took up some university study, but mostly becoming a merchant was through an apprenticeship with some company or firm. With the internationalization of commerce there was, however, an increasing number of rules and of intricate business problems. A number of encyclopedic ventures throughout Europe tried to measure up this complexity. The best known among these was Jacques Savary senior with his *Le parfait négociant* (The Perfect Tradesman, 1675) and the *Parères* (1688) that discuss difficult international trading issues. His son Jacques Savary des Brûlons later added his *Dictionnaire universel de commerce* (General dictionary of commerce, 1723). These encyclopedic books from the Savary's were translated and extended for other regions.

The first business schools were founded at the crossroads of these encyclopedic works and the old writing and arithmetic schools. The oldest one, the Aula do Comércio, was founded in Lisbon in 1759 during the Marquis de Pombal's modernization of Portugal after the earthquake in Lisbon (1755). In Hamburg, the old tradition of the *Rechenmeister* (arithmeticians) encountered modern-day trade with the *Handelsschule* that was founded in 1767 by Johann Georg Büsch. In Berlin, Johann Schulz founded a similar school in 1791. France only followed later with its École spéciale de commerce et d'industrie in 1819. All of these schools emphasized real-life situations of commerce and included not only the classic accounting, bookkeeping and arithmetic in the curriculum, illustrated with many numerically calculated examples, but now also offered a general introduction into mathematics, even including some algebra.

Surveying

A similar evolution can be observed in the surveyor trade. There existed no formal training or schooling for surveyors in the eighteenth century. Most of the surveyors came to the trade by apprenticeship, though quite some surveyors were former officers and had attended some military school. As early as 1600, the Netherlands had tried to establish a common curriculum through an engineer school, the School voor Nederduytsche Mathematique in Leiden following Simon Stevin's plan. The school was later integrated into the Leiden university. In France, the military surveyors, the *ingénieurs-géographes*, were formed at the École des ponts et chaussées (1747) or as apprentices to the astronomer family Cassini of the Academy. Those schools provided a more advanced mathematical level, Stevin's decimal fractions in Leiden or Cassini's triangulation schemes in Paris, but reached only a top level of the (mostly military) surveyors. Locally, apprenticeships and adherence to well-tried techniques and to local habits and measures persisted.

FIGURE 2.1. Frontispiece of J. G. Zollmann's *Anleitung zur praktischen Geodäsie* (1745). It depicts the surveyor in the field with his surveyor's chain and his round plane table. A Minerva-like goddess, the incarnation of science, is giving him pointers on how to apply science in his practice. Photo courtesy of ETH-Bibliothek Zürich, Rar 8885, Public Domain.

FIGURE 2.2. Frontispiece of Marinoni's Marinoni's *De re ichnographica* (1751). It depicts the learned geometer sitting in his study, surrounded by books, regular geometric volumes, and instruments. While studying the book on his desk, compass in hand, a goddess takes him by the hand and invites him onto the field to apply his knowledge. Two cupids bearing the surveyor's chain are already marching to the field. Photo courtesy of ETH-Bibliothek Zürich, Rar 771. Public Domain.

Not only did two "classes" of surveyors exist, but also two kinds of maps. On the local level, maps were "relative," expressed using local measures, detailed only where needed, and drawn up relative to landmarks structuring the landscape, such as church towers, crossroads, rivers, etc. On the national or global level, the points of reference had to be more precise, astronomical if possible, the detail systematical, and the scale and method consistent throughout to make them compatible to other maps. The latter category of more "professional" maps were often executed by military surveyors only and guarded as a military secret. In the eighteenth century, the ascent of commercial cartographic publishers serving the public interest in maps would introduce "professional" standards in the open.

Jean-Dominique Cassini, founder of the Cassini astronomer dynasty, was one of the first astronomers that had been able to determine latitude and longitude of a place, using the moon eclipses at first and later the satellites of Jupiter for finding longitude. Guillaume Delisle would use Cassini's idea of a place's astronomical coordinates as the pinpoints of a grid for drawing up his high-precision maps, all according to the same scales and measures. Delisle would set new standards for mapmaking in the eighteenth century, not only through his scientific methods and his improved equidistant conic projection, but also through his systematic and meticulous collection and analysis of cartographic and topographic information sources to achieve well-documented maps. His

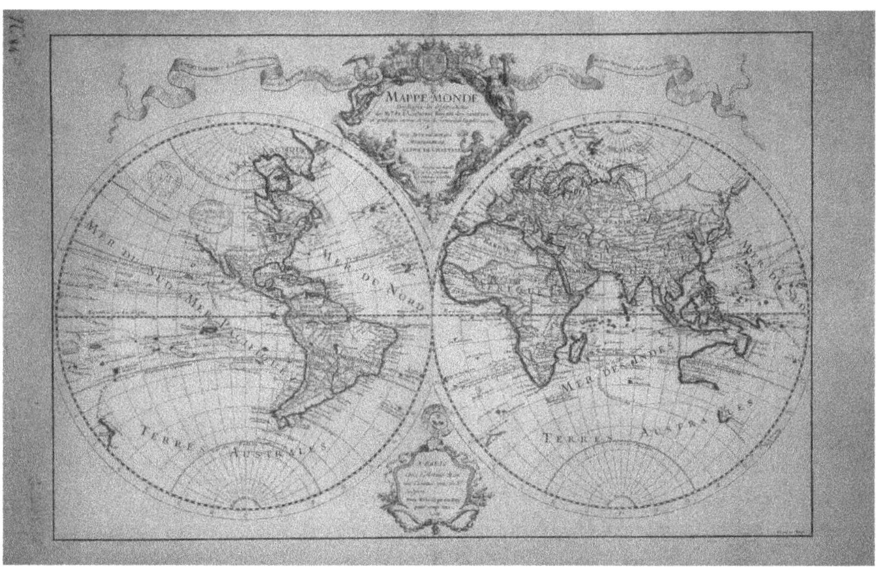

FIGURE 2.3. Guillaume Delisle's Mappemonde, map of the world (1700). Photo by DEA Picture Library / De Agostini / Getty Images.

Mappe-Monde (1700) brought new precision to maps, showing, among other things, that the Mediterranean was 300 miles shorter than previously assumed, thus presenting "a new earth" as it were, according to Fontelle.

This new, more scientific method of drawing large maps was also used by commercial cartographers. The Homann publishing house of Nuremberg early on tried to emulate Delisle's quality and asked Wittenberg professor Johann Matthias Hase to help them. Hase was equally meticulous in his cartographic and historical research, but proposed another, stereographic projection for the maps. Later still, Tobias Mayer would work for Homann during five years (1746–51), starting from astronomical coordinates he had determined himself, Mayer would help integrate the existing, local maps into larger and more accurate maps.

This did not rule out both social and practical tensions between local and global map making. Such discussions came up in the production of Russia's map. While Joseph Nicolas Delisle, Guillaume's half-brother, wanted to apply Cassini's method of topography, with a predefined triangulation network of astronomically determined positions, Ivan Kirilov argued against it. Kirilov had been in charge of the map since 1720 and was more in touch with the practical reality in far-off regions of Russia. According to him, local measurements using rivers and geographical markers as points of reference were good enough and could later be integrated and adjusted with the help of the longitude and latitude determinations of some points of the measurement. Delisle would have preferred retraining the surveyors altogether. The discussion was never fully resolved, but the eventual atlas published in 1745 actually vindicated Kirilov's point (Edney and Pedley 2019: 76–8).

The atlas of Russia would also provoke a discussion on the best projection to be used, either Hase's stereographic projection or Delisle's conic projection. This indirectly prompted Johann Heinrich Lambert's at the Berlin Academy to systematically analyze how the surface of the earth, taken as an ideal sphere, may be projected onto a flat map. Lambert took a point on the surface of the earth and projected a second point of the surface, through the calculation of the intersection of the straight line of both points with an equatorial plane, he could show how coordinates a and b of the surface, defined by longitude and latitude circles, are mapped onto "Cartesian" coordinates x and y on the plane. This led to a set of partial differential equations that Lambert analyzed in order to find out what projection preserved areas, angles, or both (the latter turned out to be impossible). This approach was later studied further by Euler, Lagrange and Gauss and had an important influence on the development of differential geometry in the nineteenth century (see Chapter 3 in Volume 5).

Increasingly during eighteenth century, rules and laws were issued that translated the state's interest in accurate and coherent mapping of the land for both taxation and planning. In France, an edict of 1669 regulated both the office of surveyor and its practices, inciting him to use Parisian *perche* as standard measure, similar edicts were issued in Prussia (1702–4) using the Prussian *Rute* as standard measure. This process accelerated after the Seven Years' War. On the one hand, in Central Europe, they had been confronted with a French army that used superior maps made by their *ingénieurs-géographes*. On the other hand, the near bankruptcy of many countries after the war pushed them to reconsider their fiscal system and reassess the cadaster. Also, in Britain, the Inclosure Act of 1773 and the Inclosure Consolidation Act of 1801 not only enabled big landowners to seize on the commons, but also obliged the local authorities to survey the grounds and have a detailed admeasurement plan. Those political and economic developments had an impact on the surveyor's profession.

The profession had to scale up its activities and in the course of this, adopt stricter procedure and use more mathematics. Instead of a surveyor and his apprentice measuring up the field with their poles and chains, larger and more professional groups had to coordinate the work. French military mapmaking mobilized a group of 7 *ingénieurs-géographes*, 3 draughtsmen, 4 plane tables, 4

FIGURE 2.4. 100 foot chain made by Jesse Ramsden for use in the Primary Triangulation of Great Britain, below is a plan of this survey conducted by General William Roy of the Royal Engineers in 1784. Photo by SSPL / Getty Images.

measuring chains and 3 or 4 soldiers to assist them. When the young United States wanted to survey the Seven Ranges in 1785, it sent out 9 surveyors with a crew of 30 chainmen and axemen. As surveyors now had to produce maps stretching over more than one locality, they had to use common measuring procedures and use more advanced techniques such as triangulation. Especially around 1800, ever more detailed regulations were issued, defining procedures on the field and standards for the official forms and maps. All amounted to the production of commensurable maps with predefined scales and templates that could, at least in theory, be easily assembled by the central administrations. This regulation from above did clash with the practice from below, generating quite some discussion.

The tension between local practice versus global theory also often touched upon the role and the amount of mathematics in practical surveying. This went into two directions, on the one hand, more and more advanced mathematics was pressed upon the surveyors, on the other hand, public servants and other people not trained in the surveyors' trade had to be able to use and interpret the new maps. For this latter category, more manuals started to appear in the second half of the eighteenth century that wanted to teach practical geometry without first going through definitions and axioms *à la* Euclid. They addressed a need for doing or using geometry in the field, without going through the university-style format of theoretical geometry. An exponent of this is Alexis Clairaut in his *Elémens de Géométrie* (1741) that tried to follow a genetic or even problem-oriented approach to geometry, to appeal more to intuition and practice. Other practical geometries started not from definition, but from the actual measuring instruments itself. The audience addressed was large, as the subtitle of Lukas Voch's book on practical geometry (1780) showed, *for the utility of public servants, lawyers, surveyors, law clerks and every other person.*

At the other end of the spectrum was the question, did the surveyor need more advanced mathematics in his trade, such as decimal fractions, logarithms, trigonometric functions, or even algebra or analysis? Many supported Christian Wolff's assessment in a preface to a treatise on practical geometry by Johann Wilhem Zollmann (1749): "One only needs as much theory as needed for the application of one's art. On the contrary, nobody can find by reason that which only through experience can be discovered" (Zollmann 1744: Vorrede). The surveyor thus only needed a minimal mathematics to do his job, though he could extend his mathematical luggage to include logarithms, trigonometric functions or a bit of algebra if advantageous (see also Chapter 1). This view was increasingly challenged as time went by. With the big triangulation networks and the increased standardization and precision that came to be mandatory by law, more advanced trigonometry became useful. Even the errors occurring in measurement, both on paper and in the field, were becoming amenable to mathematical analysis.

ACADEMIES OF SCIENCE

A rather new and typical institutional creation of the eighteenth century are the academies of sciences The foundation of the earliest learned societies and academies happened in the seventeenth century: Accademia della Crusca in Florence (1573), Accademia dei Ricovrati in Padua (1599), Accademia dei Lincei in Rome (1603), the Royal Society in London (1662), and the Académie Royale des Sciences in Paris (1666) (cf. Chapter 2 in Volume 3). Most of these started off as a rather informal gathering of literate people in the form of a learned society (*société savante*), but some of them became institutionalized. Once an official institute, empowered by government, membership of an academy became equated with official recognition of scientific competence. As a consequence, the opinions of members acquired normative value that, if needed, could be enforced. Further, their members were also deemed to use their knowledge and competence to contribute practically to the improvement of the state and its bodies. The new institutional concept flourished in the eighteenth century as is witnessed by a long list of newly found academies throughout Europe. Berlin (1700), Lyon (1700), Uppsala (1710), Bordeaux (1712), Bologna (1714), St. Petersburg (1724), Stockholm (1734), Göttingen (1751), Erfurt (1754), Turin (1757), München (1759), Mannheim (1763), Edinburgh (1784), Prague (1784), Dublin (1785), etc. This list is far from exhaustive, but it shows the academy could be found both in the capitals of Europe as in the smaller (university) cities.

The academies, especially the larger and state-funded ones, were institutions that awarded a limited number of pensions to some of the most prominent scientific minds in all branches of the sciences, ranging from the *belles-lettres* (history and literature) to the exact sciences. Though each academy had its own structure and particularities, all academies had a hierarchy among its membership. Slightly generalizing from the particular cases, there were about three main categories of academicians: the full members, the associate members and the corresponding and/or foreign members. One had to ascend through the ranks from the lower to the higher categories, with only the full members receiving a full pension, the other positions being stepping stones towards the full pension. In most academies, membership did not suffice for a living and had to be supplemented with other sources and opportunities of income, or rather vice versa, the academy's pecuniary contribution in most cases was just a supplement. Most of the academicians had a job within the government, many in the educational system (the *collèges*, the universities, the officer's schools, etc.), but frequently they also held a consulting or supervisory function in more technical endeavors such as the Mint, the mines, the navy, the military, the observatory, etc. (Hahn 1975). Next to their contributions to governmental matters, the academicians also played an important part in the Republic of

Letters, the eighteenth-century public space of opinion that was constituted of letter correspondences, journals, pamphlets, and books.

Both the mission and the importance of academies differed greatly among each other and depended foremost on the individuals working at the academy, although the interest of the sponsoring institution could also have an important influence. As far as mathematics was concerned, the most important academies during the eighteenth century were surely those of Paris, Berlin and St. Petersburg. The Royal Society, after Newton's presidency, contributed rarely to the mathematical discourse, focusing instead on experimental science, the antiquarian and the curious. In the United Kingdom, it was rather the new military academy in Woolwich or the Scottish universities that would be the places where mathematics would thrive (see the next section). Whereas the Royal Society received but little state (financial) support and only in the late eighteenth century was called upon as a consulting instance for the government, the academies of Paris, Berlin, and St. Petersburg were mainly state-financed and, hence, instruments of the state.

The Académie Royale in Paris had an important role in the sanctioning of royal privileges (see also Hahn 1971). Scientific and technical books had to pass the academy before getting the privilege to be printed, inventions had to be studied by the academy before getting royal privilege (or not). In France, the academy was thus the ultimate instance in matters scientific. Sending publications or inventions to the academy for their approval was at the same time displaying the aspiration to become associated with the institution. Its class system of adjoint, associate and full members assured the interaction with (and partial control of) the Parisian learned circles. Professorships at the better *collèges* or at the military schools were rewarded with associate membership, members of the royal court (counselors, doctors, royal tutors, etc.) could obtain membership, etc. Similar structures of membership were at work in the academies of Berlin and St. Petersburg.

The normative or regulatory function of the academies did not exclude discussions and quarrels within the academy or among academies, though it was considered bad practice that these discussions went public. Of course, some quarrels, such as the priority debate over the calculus (see Chapter 3) or the Pierre Louis Maupertuis–Johann Samuel Koenig fight over the principle of least action (see Chapter 5), went public and attracted much interest. Most discussions, however, did not leave the academy and could only be tracked down in the notes of the academy and the letters of its members. Sometimes, rumors and allusions got printed in the popular gossipy memoirs of the time. In many cases, the discussions were not only on scientific merit or priority, but were symptoms of infighting between rivaling factions within the academy, trying to win a struggle over power, influence, and appointments.

An instructive case of such internal issues is that of the former Jesuit Rudiger Boscovich at the Paris Academy. After the dissolution of the Jesuit order (1773), Boscovich had obtained a professorship at a French naval school and now tried to secure a position at the academy. He was helped by the academicians Charles Marie de la Condamine, Clairaut, and Lalande, but was obstructed by a rivaling faction, led by Jean le Rond d'Alembert. Boscovich tried to garner scientific attention with numerous scientific memoirs and inventions, one of which was a proposed improvement of a micrometer. However, this improvement clashed with a previous invention by Alexis Marie Rochon already validated by the academy a few years earlier. Reluctantly, Boscovich conceded that Rochon had read to the Academy on the micrometer before him. But he still maintained his (independent) priority claim and sent an article on his invention to the Royal Society where it was printed in its *Philosophical Transactions* of 1777. By doing so, Boscovich had bypassed the Paris Academy and dragged the issue in public. As a result, Boscovich lost much credibility and goodwill at the Academy. Seeing his efforts frustrated to obtain a better pension, Boscovich eventually returned to Italy in 1783.

Some of the most renowned mathematicians of the eighteenth century were academicians. Leonhard Euler first became member of the Saint Petersburg Academy (1727–41), than changed to the Berlin academy (1741–66), but returned to Russia (1766–83) after frictions with the Prussian King Friedrich II and other academy members. Joseph-Louis Lagrange first worked at the academy of Turin (1757), before being called to Berlin as Euler's successor (1766). He was finally made full member of the academy in Paris (1787) and would retain his scientific status after the French Revolution during the First Republic (1792–1804). With abolishment of the academy, Lagrange would now teach at the newly founded institutions: the École normale and the École polytechnique (1794). Johann Heinrich Lambert after touring the universities and academies with his pupils became founding member of the Academy of Munich (1759), before joining the Academy of Berlin in 1764. Because of the pyramidal structure of the French academic system, there was less geographical mobility, but throughout the eighteenth century the Academy of Paris could pride itself of a number of important mathematicians: Michel Rolle (1685), Pierre Varignon (1688), Thomas Fantet de Lagny (1696), Bernard de Fontenelle (1697), Jacques Ozanam (1702), Nicolas Guisnée (1705), Alexis Clairaut (1732), Jean le Rond d'Alembert (1741), Étienne Bézout (1758), Alexandre-Théophile Vandermonde (1771), Nicolas marquis de Condorcet (1769), Gaspard Monge (1780), and Adrien-Marie Legendre (1783). This list gives their year of entry in the Académie, when they became members *adjoint*. This was mostly when they were in their late twenties, early thirties, it would take them often some twenty years more before they could become full member.

FIGURE 2.5. *King Louis XIV visits the Paris Academy* (1671), engraving by Sébastien Leclerc the Younger. Photo by Heritage Art / Heritage Images / Getty Images.

Some eighteenth century women also rose to scientific prominence, they were mostly of noble descent, though, in other cases, they were married to a member of an academy or university. It was crucial to have access to salons, libraries and correspondence networks. In general, women could not become members of an academy, with the sole exception of the Academy of Bologna. The academy, under protection of pope Benedictus XIV, had elected the mathematicians Laura Bassi and Faustina Pignatella as (honorary) members in 1732. Bologna later also elected Émilie du Châtelet, known for her French translation of Newton's *Principia*, and Maria Gaetan Agnesi, who wrote one of the first manuals on the calculus (1748), as members.

While many of the smaller and local academies were rather a haphazard collection of individuals with no precise research objectives in the long run, the bigger academies invested into bigger projects. Some of these projects required an investment in time and instruments and the guarantee of continuity over time, such as astronomical observations in an observatory, cartographic projects, or meteorological observations. Others demanded an organizational investment, dispatching academy members with personnel support into the provinces of the state, or even into faraway regions of the colonized world. Some of the best financed long-term projects performed by the academies were to obtain detailed maps of the territory of the state. The Cassini family of astronomers and surveyors made a complete geodetic grid for all of France in 1744 and continued to dress up detailed maps for all regions of France from 1744 to 1783, in the meantime noting down the number of births and deaths found in the church registers to estimate the population. In Russia, the Saint Petersburg Academy had a long running project to produce an atlas for all of Russia, including parts that had been scarcely charted before, eventually the atlas appeared in 1745. The most spectacular of the larger academy projects were, however, the big expeditions in the early eighteenth century to assess the shape of the earth and the later expeditions to gather and collect all sorts of data (astronomical, geographical, botanical, zoological, etc.) in various parts of the world. Those expeditions also appealed to a general public that perused the travel accounts and the descriptions of the foreign and exotic. The expeditions to measure up the earth run through the eighteenth century and were mostly set up by members of the Paris Academy. They include both the measurement of parts of the meridian at home (Paris, Rome, Vienna), and abroad, such as the expeditions of Pierre Louis Maupertuis and Clairaut to Lapland (1733–7), of la Condamine, Pierre Bouguer and Louis Godin to Peru (1735–43), of Nicolas Louis de Lacaille to the Cape of Good Hope in 1751, or of Charles Mason and Jeremiah Dixon in North America (1763–7). The results of Maupertuis' expedition vindicated that the earth was flattened at the poles and gave him a wide notoriety throughout Europe. Later in the eighteenth century, in 1787, a final measurement of the distance between the observatory in Greenwich

FIGURE 2.6. Portrait of Pierre-Louis de Maupertuis showing how he “flattened” the earth, below is depicted a scene from his Lappland expedition. Photo by DeAgostini / Getty Images.

and the observatory in Paris provided the missing part for stitching up the two global grids that the two upcoming colonial powers of the eighteenth century, France and Great Britain, had established. This provided the occasion to propose a universal unit of distance to be used by all astronomers and geographers, an idea that would later result in the metric system (see the Introduction).

Many other researchers at the academy were stimulated by the immediate needs of the royal or imperial bodies. First and foremost problems of military nature: ballistics, military cartography, shipbuilding and others. But also problems of infrastructure and industry: architecture, canals, wind- and watermills, navigation at sea, chemistry. And finally, some activities that provided income for the state and the academy: lotteries, ephemerids and calendars. All of those problems led to scientific articles in the academies' publication venues, though often the practical origin and the royal demand behind the scientific question retreated in the background or could even be obscured because it was considered a state secret (such as precision maps or ballistic tables).

Also scarcely noticeable to the public eye was the organization of labor behind academic work. While one (senior) name was on the paper, it was often the result of combined forces, often unacknowledged in print. The case was most obvious for the observatories. The head astronomer and his assistant, both of them member of the academy (as full and associate/adjoint, respectively), were mentioned, but much of the actual observations and the main calculations on those observations would be done by the assistant. When many, repetitive calculations had to be made, one would often enlist the help of non-academic personnel. In some cases, the astronomers' family members, often female, were helping out, in other cases, male and female calculators were paid by the amount of work done, the work could be performed at home or in the astronomer's office. A similar division of labor could be found in ballistics, where the drudgery of computing tables was left to a junior officer but the publication of the theory with its tables to the academy member. Secretaries and writers also helped alleviate the academician's work. The old and nearly blind Euler even had a number of assistants that read him his correspondence in the morning, than took down in writing the ideas for academic memoirs Euler wanted to write. In the afternoon, they came back with the memoir all written out, calculations and derivations included, for inspection by the master. There were also cases where the academicians teamed up with amateurs to further labor-intensive studies, instances are the exchange of meteorological data, or Lambert's project to produce number-theoretical tables (1770). In these cases, the private correspondence network served as the organizing structure for collaborative research.

FIGURE 2.7. Frontispiece of the first publication of the Berlin Academy, the *Miscellanea Berolinensa* (1710). The Athena-like figure in the middle is surrounded by the instruments of the sciences. The two cupids represent the two sides of writing: the Belles Lettres on the left side, mathematics on the right side. Note that the mathematics book holds a differential equation and within the circle the Leibniz series for *pi*. Photo courtesy of Berlin-Brandenburg Academy of Sciences and Humanities, Public Domain.

UNIVERSITIES AND SPECIAL SCHOOLS

Schools reproduce and transmit mathematical knowledge and through that reproduction have an important impact on what and how mathematics is transmitted into society. On top of that, some of the institutes for higher education also play a part in the production of mathematical knowledge and know-how. One format of higher education were the universities, an institution that had come down from the late Middle Ages, further there were the preparatory and special schools, such as military or "technical" schools. During the eighteenth century, there was also a growing state involvement in the foundation and organization of these schools (cf. Chapter 1 for elementary schools).

Universities

Although the university was an institutional creation from the late Middle Ages (thirteenth and fourteenth centuries), it had proliferated all over Europe between the fifteenth century and seventeenth centuries. The eighteenth century mainly inherited these institutions together with their laws and structures, though a few, so-called "reform," universities were newly founded (cf. de Ridder-Symoens 1996). One of the innovations of these new institutes was the use of the vernacular instead of Latin as the language of teaching. The main function of the university, preparing for a career in law, medicine or religion, remained unchanged for most universities in the eighteenth century. In other words, eighteenth-century universities were, save a few exceptions, not places of research, but of tradition and classic formation. The professors mostly did not engage in research activities, but just dispensed knowledge tested by time and approved by the authorities. Although the curriculum was slightly updated, the general structure of a preparatory (philosophical) faculty followed by a specialized faculty was kept throughout. Mathematics was part of the philosophical faculty, where music theory, arithmetic, and geometry had their traditional place. By the eighteenth century, this had become in most universities an encyclopedia-like introduction into the elementary parts of mathematics. This comprised an introduction to the bibliography of mathematics, especially into the ancient Greek and Roman sources, some elementary arithmetic in the decimal positional system, and finally, an often simplified version of Euclid's Elements introducing the students to geometry. The newer developments in mathematics, the parts we now call algebra and calculus, were absent in many if not most universities. If some elements of the newer mathematics were taught at all, this was not during the public lectures, that all students enrolled at the university could attend, but rather it was treated in the *privatissime*, the private courses professors offered against a fee.

Through Leibniz's diplomacy, the University of Padua in Italy would be one of the first places were the infinitesimal calculus would be taught. First Jakob Hermann, later Nicolaus I Bernoulli would teach in Padua (1707–13 and 1716–22, respectively) bringing the Leibnizian calculus to Italy early on. In the German protestant states, the reform universities of Halle and Göttingen would be among the earliest to offer some entry into more advanced mathematics. Probably one of the most influential figures in this respect is Christian Wolff. With the patronage of Leibniz, Wolff had become professor in 1706 at the newly founded (Prussian) University of Halle (1694). After a dispute with the Pietists, who stood for a strict and dogmatic form of Protestantism, Wolff was banned from Prussia and took up a professorship at Marburg (1723), returning to Halle only in 1740. His first years in Halle were marked by a prolific output of manuals, especially on mathematics. He first published his *Anfangsgründe der mathematischen Wissenschaften* (1710, *Elements of the Mathematical Sciences*) in four volumes, developing the German mathematical terminology along the way. The fourth volume contained an introduction to algebra and the infinitesimal calculus, supplemented with a bibliography of mathematics. Wolff than subsequently published a Latin version of his volumes, *Elementa matheseos universae* (1713ff., 6 vols), that contained more details and proofs, as well as a one-volume abridged version in German meant as the basis for a mathematics course at the universities, the *Auszug aus den mathematischen Wissenschaften* (Excerpt of the Mathematical Sciences, 1713). Finally, he also published his *Mathematisches Lexicon* (Mathematical Lexicon, 1716), a dictionary of mathematical terminology. While, except for the *Auszug*, they were mostly only taught in private courses, these books and their many re-editions became the de facto reference books for two generations of mathematically interested students and professors at many German universities. They also exerted a considerable influence outside the German-speaking countries, in France, Italy, the Scandinavian countries, Holland, or Russia. Even though algebra and the calculus were only briefly treated, their inclusion in these works and their systematic exposition brought them to the attention of university students. The addition of a commentary on the mathematical literature also gave them further pointers for self-study.

In the latter third of the eighteenth century, a number of German authors provided more modern versions of Wolff's textbooks. They essentially stuck to the combination of an encyclopedic introduction to mathematics with notes and comments on the newer mathematics and mathematical literature. One of the prominent figures was Gotthelf Abraham Kästner, professor of mathematics at the new University of Göttingen. The Duchy of Brunswick-Lüneburg, that was dynastically connected with the then ruling house of Hannover in Great Britain, had founded the University of Göttingen in 1734. Though situated in a rather provincial town, Göttingen became a gateway for scientific information, especially between the German states, Great Britain and the Scandinavian countries. Albrecht von Haller was pivotal in this, especially through his creation of the biweekly journal *Göttingische Gelehrte Anzeigen* (Göttingen News from the Learned World). The *Anzeigen* reviewed new books and had an important, even normative, impact on the scientific landscape in Germany. Further, an academy was added to the university, making Göttingen an early example of some kind of "research" university.

Kästner fitted this Göttingen profile, he was not so much a creative mathematician, but rather a very active information broker with an enormous correspondence network. He was responsible for nearly all German reviews of mathematical books in the second half of the eighteenth century, serving as the authority on mathematical publications for the German public. As a result, his rich library received the newest publications for review and thus became the most up to date mathematical library in the German speaking countries. His textbooks also drew heavily on that literary treasure. His *Anfangsgründe der Mathematik* (Elements of Mathematics, 4 vols, 1758–69) would become more popular than Wolff's in the later eighteenth century, its first and second volume were often used for public lectures on mathematics. Kästner later published his *Geschichte der Mathematik* (History of Mathematics, 4 vols, 1796–1800) supplementing his textbooks. It contains a history of mathematics that is mainly a huge collection of chronologically and thematically arranged quotes from his extensive library of historical and contemporary mathematics books.

The same divide between "classic" parts of university mathematics (essentially elementary arithmetic and some Euclid) and more advanced parts such as algebra and the calculus can also be observed in other countries. In the Netherlands the university of Leiden had a venerable tradition of teaching more advanced mathematics since the seventeenth century. In the early eighteenth century, Newtonian William Jacob 's Gravesande was responsible for mathematics. He published a *Physicis elementa mathematica, experimentis confirmata, sive introductio ad philosophiam Newtonianam* (Mathematical Elements of Physics, Confirmed by Experiments, or an Introduction to Newtonian Philosophy, 1720) and a *Matheseos universalis elementa* (General elements of mathematics, 1727) for his lectures in physics and mathematics. While the latter book contained some introduction to Newton's natural philosophy and the former an appendix on algebra commenting Newton's *Arithmetica Universalis* (1707), the public lectures mainly treated the "classic" encyclopaedic parts. Again, more advanced topics, if taught at all, were reserved for private tutoring.

Parts of 's Gravesande's manuals would in their turn be included in the manuals that Abbé Nicolas Louis de Lacaille wrote for his students at the Collège des Quatre Nations. Once founded by the Cardinal Mazarin (1686), the school was one of the most prestigious schools in Paris and part of the university with a chair of mathematics. The chair's first incumbent was Pierre Varignon, one of the promoters of the new infinitesimal calculus who had worked intensively on applying it systematically to statics and mechanics (see Chapter 3), though little or nothing of this work appeared in his teaching. When de Lacaille became mathematics professor, he started using French as the language of teaching instead of Latin and published his manuals, the *Lecons élémentaires de mathématiques* (Elementary Lessons of Mathematics, 1741) and the *Lecons élémentaires d'astronomie* (Elementary Lessons of Astronomy, 1746). Later editions would incorporate much of the newer mathematics, especially once his successor abbé Joseph-Francois Marie took over both the teaching and the manuals. De Lacaille's manuals explicitly differentiated between parts of the book that were part of obligatory curriculum, and parts that were optional, viz. for private courses or for self-study. Using a typographical distinction, already employed by Wolff, the optional parts were set in small letters, "those want only to know the first elements of mathematics, can skip all passages in small letters," however, those who wanted to read mathematical books for themselves were urged to "study everything carefully." The textbooks were re-edited several times until the end of the eighteenth century and were also translated into Latin by the Jesuits. In the Oratorian Schools of France, the manual of Guisnée, *Application de l'algèbre à la géométrie* (1705), was used, showing how algebra could be applied to geometry.

Also the mathematical and physical curriculum of the important Jesuit schools was updated during the eighteenth century. The prohibition of teaching the Copernican (or for that matter, Newtonian) heliocentric physics was first softened, a heliocentric system could now be proposed as a hypothesis, before being lifted completely (in 1757). The opposition against indivisibles was also slowly abandoned. More even, quite some Jesuits were occupied with (Latin) translations of textbooks, such as de Lacaille's or Colin MacLaurin's book on fluxions. In practice, just as most universities, their schools remained focused on geometry, elementary arithmetic, and plane trigonometry sometimes including some simple algebra. From 1740 onwards, some intimations of infinitesimal calculus could also reach the curriculum. A symptom of this "thawing" up to the calculus is the fact that the chair of mathematics at the Collegio Romano was held by Rüdiger Boscovich (from 1740 to 1750) and later his student Carlo Benvenuti (from 1750 to 1752). Boscovich would develop an intricate, unified synthesis of Cartesian, Newtonian and Leibnizian worldviews in his *Theoria philosophia naturalis* (Philosophical Theory of Nature, 1758, second revised edition, 1763) and had promised, though did eventually not publish, a volume on differentials and integrals in his *Elementorum Universae Matheseos* (Elements of General Mathematics, 1754) textbook for the Collegio. The matter remained a touchy topic for Jesuits and one had to remain cautious, as the eventual removal of Benvenuti to another post showed, when Benvenuti became too vocal in saying the heliocentric system was not hypothesis but reality.

In the United Kingdom, most universities stuck to elementary arithmetic, Euclid's *Elements* and some practical geometry too. For most students at Oxford or Cambridge, mathematics was elementary arithmetic and the first two books of Euclid, without demonstrations, in Robert Simson's translation or one of its abridged editions. Elementary algebra was sometimes advertised in the more advanced courses, and some professors did even introduce lectures on Newton's fluxions in the mid-eighteenth century. The blind mathematician Nicolas Saunderson introduced them at Cambridge, Robert Simson at Glasgow, James Gregory and Colin MacLaurin at Edinburgh (Guicciardini 1989).

Although there is certainly a slow spread of newer mathematical topics at a number of universities during the eighteenth century, it must be kept in mind that, in practice, these courses were nearly almost *privatissime*. If not enough students appeared (and paid the extra fee), the private course was simply canceled. The public lectures only featured the general, encyclopedic introduction to mathematics as part of the official curriculum of the philosophical faculty. People looking for an in-depth introduction to the newest mathematics—the infinitesimal calculus of Newton, of Leibniz, of the Bernoullis—did not find it at the universities. Instead they had to turn to the few conversant in the new mathematics for private courses, for example, Johann Bernoulli's private courses to Guillaume de l'Hôpital that resulted in l'Hôpital publishing *L'Analyse des infiniment petits pour l'intelligence des lignes courbes* (Analysis of Infinitesimals to Study Curved Lines, 1696). More often, they could read about it themselves, and as an autodidact, could educate themselves. This, however, still depended on access to the published books and memoirs. This was limited to those who either had the means to buy books, or to those who could borrow volumes either in an institutional library (e.g., of a university) or in a personal library to peruse, copy, and annotate.

Other Institutes of Higher Education

Around the Austrian Succession War (1740–8) and the Seven Years' War (1756–63) many states founded new military schools. Sobered up by the wars, England, France, several German states and other nations wanted to modernize the education of their officers. This led to either the modernization of existing military schools or the foundation of new schools, often for engineers and for artillery officers, all including more mathematics in the curriculum. Russia founded its Military Academy in St. Petersburg in 1717, commanding Euler to write the manual for arithmetic. Great Britain founded its Royal Academy in Woolwich in 1741, France its École royale du génie de Mézières in 1748 and the new École militaire for impoverished noblemen in 1750, Vienna its Royal

Academy in 1751. Whereas nobility was often the sole criterion for entering a military school before the mid eighteenth century, many nations now allowed ordinary young men to become officers after passing an exam. This did not happen without resistance, the French aristocracy successfully blocked changing the entrance criteria of the traditional naval schools that only admitted young men that were of noble descent for at least four generations straight. In 1775–6, they also managed to abolish the new system that let ordinary men study at the École royale du génie de Mézières (since 1762) and the École royale de la Marine du Havre (since 1773). While in other countries the rules for entry to officer schools were often less strict than in France, in reality, most of their recruits also descended from the better classes. The democratization of the schools was only successfully pursued after the French Revolution.

In France, from 1764 onwards, a centralization of the mathematics curriculum of the existing naval and military schools was enforced by the appointment of an *examinateur*, a mathematics professor who was tasked with writing manuals and examining the officers in training. For the artillery and engineer schools, the academician Charles Camus was appointed, for the naval schools, Étienne Bézout. The latter's *Cours de mathématiques à l'usage des Gardes du Pavillon et de la Marine* (Mathematics Course for the use of the Admiral Guards and the Marine, 1764–9), five volumes on arithmetic, algebra, geometry, calculus, mechanics, and naval sciences, would become a success, despite its inclusion of quite some advanced mathematics. After the death of Camus in 1768, Bézout also became *examinateur* for the artillery schools and adapted his manuals accordingly. Charles Bossut succeeded to Camus' function as *examinateur* in the engineering schools and, following Bézout's example, published his own manuals. When Bézout retired in 1783, he would be succeeded by Gaspard Monge for the naval schools, and by Pierre-Simon de Laplace for the artillery schools. After the French Revolution, Monge and Laplace would become professors at the newly established Écoles normales and the École polytechnique (1794) (cf. Chapter 2 of Volume 5). This shows how, in spite of the revolution, there was much continuity between the reform of the military schools in the 1760s and the establishment of the new schools of higher education. Also, the manuals by Bézout, Bossut and de Lacaille would remain popular in secondary schools even at the beginning of the nineteenth century, when the manuals written by Sylvestre Francois Lacroix would slowly replace them.

The French example of military schools was often copied, some educational "entrepreneurs" even opened so-called "French schools" abroad. The kingdom of Hannover sent out the Göttingen professor A. L. F. Meister on a trip to all military schools in France and published a report afterwards (1765). Stuttgart founded its Hohe Karls-Schule in 1770 after the French model, Berlin tried to do the same, asking the Academy of Sciences to cover for advanced courses and translating into German an artillery manual from the military school in Turin (founded already in 1678!). Spain merged its two existing artillery schools in Cadiz and Barcelona in a new institute at Segovia (1762–4) and later also founded a school for naval engineers (1772) trying to keep up its power at sea under pressure of the new colonial powers Great Britain and France. Portugal followed with a school for naval engineers in 1782.

With the revolutionary wars with France upsetting the political equilibrium both in Europe and the colonies at the end of the eighteenth century, new military academies popped up everywhere: Berlin finally got its school for engineers (1788) and artillery officers (1791), similar schools were opened in Hannover (1782), Naples (1787), Holland (1789), Dresden (1796), etc. Great Britain opened its Royal Military College in Sandhurst in 1799. The military schools also spread beyond Europe. After the example of Lisbon (1790), a military academy was opened in Rio de Janeiro in 1792. The United States started its academy at Westpoint in 1802.

These new schools often generated discontent and discussion. As mentioned before, instead of the feudal principle of entitlement by descent, a selection of military personnel (and of civil servants) on the basis of capacity and merit was now tried out cautiously. Also the introduction of more mathematics did not go by without debate. The age-old discussion of the alleged benefits of theory for practice and of the practical experience that theory cannot supplant was reinvigorated. The artillerist complained that ballistic theory was not useful in practice, the seamen stuck to their old practice of determining position at sea by dead-reckoning instead of embarking on simple but long calculations with lunar tables, etc. (cf. also Chapter 6) As a mark of the institutional reform and modernization, however, mathematics often won the debate, in the sense that it stayed in the curriculum, though its application in practice may have often been lacking.

THE MEDIA OF COMMUNICATION

One of the characteristics of the eighteenth century is the exponential growth of literary communication, creating the so-called "Republic of Letters." The establishment of reliable postal services since the seventeenth century, the growing circulation of books and printing and the emergence of journals made the world, or at least Europe, a connected world for the literate. Transnationally, the preferred languages were French and Latin, though nationally, the vernacular was used. This communication revolution had its impact on mathematics, too. Whereas discussion on mathematically more advanced topics had typically been confined to learned correspondences in the seventeenth centuries, these debates started to surface in the new public sphere of periodical literature during the eighteenth century.

For a large part, reading and writing were simultaneous activities in the eighteenth century. Reading a book meant writing down excerpts and taking down notes, those excerpts and notes could in their turn be used to write letters and books. Indeed, literary correspondence in the eighteenth century talked a lot about books, about their publication, their local availability, their content, and there was a lot of to and fro of asking for excerpts from books, or for borrowing books. Mathematical books, representing about 5 percent of the total literary production, were a natural part of this literary world throughout. As books were expensive and specialized books such as those in mathematics quite rare, the select persons interested in mathematics naturally sought to exchange both books, questions and ideas with the like-minded, sometimes locally in clubs, more frequently through a network of correspondents. Often during the seventeenth and the eighteenth centuries, discussions started in essentially private, learned correspondences, but later came out in the open through their publication as a separate book or through quotations in the comments, notes, and margins of new books. One may think of the correspondence editions appearing under the title *Commercium Epistolicum*, such as in the famous discussion between Wallis, Brouncker, Fermat and Frans van Schooten (1658), or between Newton and Leibniz (1713). But most of the correspondence was rather of a varied nature, collecting, quoting, reporting, and commenting on what went on in the literary world. From these private correspondences eventually sprang the first periodical publications, the learned journals.

The oldest learned journals date from the late seventeenth century, but they proliferate in the eighteenth century. In Great Britain, there were the *Philosophical Transactions of the Royal Society* (1665), initiated from Henry Oldenbourg's correspondence network; in France, the *Journal des Scavans* (1665); in Italy the *Giornale de' letterati* (1668), and in the German-speaking world the *Acta Eruditorum* (1682), born from Otto Mencke's correspondence, all published on a monthly basis. In the eighteenth century, the academies followed with the publication of their own proceedings, mostly on an annual basis. The Académie Royale in Paris started publishing its *Histoire de l'Académie royale des sciences avec les mémoires de mathématique et de physique tirés des registres de cette Académie* (History of the Royal Academy of Sciences with the Memoirs of Mathematics and Physics Taken from the Academy's Registers) in 1666. The Academy of Berlin first published the *Miscellanea Berolinensia* in Latin (1710 onwards), then changed to French with its *Histoire* (1745) and *Mémoires* (1770). In Continental Europe, Latin and French were the *linguae francae* for scientific communication. In Britain, English was mostly used, but in most Continental countries, the vernacular as a scientific language only became viable later, in the nineteenth century, though as early as 1763 the Academy of Munich decided to publish their *Abhandlungen* in German.

These early learned periodicals often retained characteristics of their origins. Instead of the self-contained essay or article, many contributions to the

Philosophical Transactions, or to the *Mémoires* of the Paris or Berlin Academies were in the form of letters written to the perpetual secretary or to one of the academy members. The *Histoire* was even more reflective of the learned life, summing up all the communications received, sent or read and discussed at the academy.

In the absence of public libraries, the correspondence network also was one of the most important ways to gather information on the Republic of Letters and their products. It need not wonder that early on rapidly appearing journals (biweekly, weekly, etc.) were published describing and discussing the newest apparitions on the book market. The *Acta Eruditorum* contained many book reviews, mostly quite objectively summarizing their content (the addition of the critical analysis is a phenomenon of the later eighteenth century). In many cities, often with the help of local university persons, *Gelehrte Anzeigen* or similarly entitled periodicals appeared with the latest news of the Republic of Letters. Others published digests of their readings, with excerpts, reviews and translations of texts having appeared elsewhere, the practical absence of copyright enabling these derivative publications. Whereas these early periodicals were started by the learned community, in the latter third of the eighteenth century, stimulated by an increased readership, literary *entrepreneurs* started publishing periodicals out of commercial interest.

Another popular format was the almanac, a yearly calendar with indication of events in the religious and civil year and of astronomical phenomena such as sunset, sundown, the moon phases or eclipses. Almanacs could contain many other things, recipes, puzzles, stories etc. Perusing an almanac was typically a pastime within the family or friend circle. Almanacs sold in the thousands of copies, much more than books that usually sold in the hundreds at best. Its revenue was thus considerable enough, and, e.g., the Berlin academy was in part financed by the sale of the monopoly on the Berlin Almanac. In Great Britain, a number of almanacs contained mathematical problems and puzzles on a regular basis. The best known is the *Ladies' Diary or Women's Almanack* (1704–1841), later its rival the *Gentlemens' Diary* (1755–1841), or, the satirical *Poor Robin's Almanack* (1663–1828), where mathematics made a more sporadic appearance (Wardhaugh 2012). Many of the mathematics appeared in the forms of riddles or questions and answers, echoing the much older tradition of recreational mathematics (see Chapter 1 in Volume 3) on the one hand, and paying tribute to the typical eighteenth century interaction with the reader on the other hand. In particular the many "philomath" societies that had popped up all over Great Britain in the 1710s and 1720s were active contributors to these almanacs, and as their interest in mathematics advanced so did the problems and riddles. Though the origins of the mathematics in these almanacs were rather in practical arithmetic and in ready reckoning examples, algebra was often used and even the calculus made regular appearances from 1730 onwards. This is reflected in the status of the editors of the *Ladies' Diary* or *Gentlemens' Diary*. At first, there were arithmetic school teachers such as John Tipper or Charles Hutton, later also teachers from the Woolwich military academy, such as Hutton, who had become appointed there in 1773, or Thomas Simpson. In other countries, mathematics did not seem to have played such a role in the almanacs, though collections of recreational problems did remain popular in new editions.

In the last decades of the eighteenth century, particularly between 1780 and 1795, a new phenomenon appeared on the literary market, specialized journals. The readership had multiplied and emancipated itself, and instead of the generalist periodicals, periodicals on particular topics for a niche public became intellectually viable, though its financial footing would prove to be less stable. They could contain original articles, extracts from correspondence, reviews, announcements, translations, and excerpts. In the German states, the first specialized mathematical journals appeared, their variety reflect the "kinds" of mathematics that were going around (Morel 2014). There was the *Arithmetisches Magazin* of J. F. Petersen (1785 and 1787) or the *Arithmetisches Vademecum* of C. J. Illing (1793), both publishing questions and solutions to simple mathematical problems, meant as exercises for those training at an arithmetics school. These were close in spirit and public to the English almanacs. There were also the journals that represented university mathematics, such as the *Mathematisches Magazin* (1803) by H. W. Kraushaar and especially the journals edited by Leipzig Professor Carl Friedrich Hindenburg.

Hindenburg first co-edited the *Leipziger Magazin zur Naturkunde, Mathematik und Oekonomie* (1781–5) that later became the *Leipziger Magazin für reine und angewandte Mathematik* (1786–9). The latter journal was co-edited with Johann (Jean) III Bernoulli who disposed of Lambert's unpublished manuscripts and also was a very active *entrepreneur* in this early scientific publishing market. Jean III Bernoulli had already published the description of his voyage along all astronomical observatories in Europe in regular episodes (1771–9), other travel accounts in Eastern Europe followed (1779–80). He had also translated Euler's *Algebra* into French upon d'Alembert's request (1774), and had tried to revive the *Acta Helvetica* of Basel (1787). The *Leipziger Magazin für reine und angewandte Mathematik* is candidate for being the first mathematical journal, though its content remains largely in the confines of the university mathematics of its time. It contained textbook level expositions of parts of mathematics and practical applications of mathematics, mostly involving arithmetic, geometry, and algebra, but rarely analysis. When Hindenburg started his final publication venture, the *Archiv für reine und angewandte Mathematik* (1795–1800) and the *Sammlung combinatorischer-analytischer Abhandlungen* (1796–1803), its contents changed. Whereas the *Archiv* pursued the earlier *Leipziger Magazin*, the *Sammlung* collected work on combinatorial analysis, a discipline which Hindenburg had founded and had now found a number of students all over the German states. The enterprise did not survive the wars with the French and had to be discontinued due to lack of subscriptions.

In other countries, early mathematical journals seemed most frequently to appear in the immediate context of a school or university. In France, the new École polytechnique published the *Journal de l'École polytechnique* (1795) with memoirs of its professors (Lagrange, Laplace, Simon Denis Poisson, etc.). In Great Britain, a number of professors of the Sandhurst College, notably Thomas Leybourn, published the *Philosophical and Mathematical Repository* (1795–1815), where the calculus and discussion and translation of French mathematics would play an important role. The members of the Analytical Society at Cambridge (Charles Babbage, John Herschel, and George Peacock) would not much later undertake a similar venture. With the translation of Lacroix' treatise on the calculus (1802, trans. 1816) and the publication of their *Memoirs* (1813) they managed to translate Continental mathematics (especially Laplace) to the Anglo-Saxon mathematical world. A proper mathematical journal that was not tied to an academy or school and was open to unsolicited contributions, even beyond national boundaries, would only appear in the early nineteenth century (cf. Chapter 2 in Volume 5).

CHAPTER THREE

Inventing Mathematics

From Solving Problems to Building Theories

JEANNE PEIFFER

In the seventeenth century, mathematics underwent profound transformations affecting methods of proof and objects of investigation. The eventual invention of the calculus, by Gottfried Wilhelm Leibniz and Isaac Newton represents the starting point in a new phase of mathematical research that led to the emergence of a new field of activity: infinitesimal analysis. This chapter is largely devoted to this outstanding invention of the eighteenth century.

In antiquity, mathematics consisted essentially of just two branches: arithmetic and geometry. With the contributions of Islamic mathematicians, later pursued and extended by sixteenth-century Italians, a third field came into being: algebra, which mainly included methods for solving equations. A further refinement came in 1591, when François Viète invented what he called a new "analytical art," i.e., symbolic algebra, a tool coupled with his explicit ambition to "solve any problem." René Descartes applied this new form of algebra to geometry, thereby considerably expanding the realm of geometric objects available to study. Descartes' fundamental idea of studying geometric objects defined as loci of points satisfying a given type of equation opened yet another door. The full potential of this conception only unfolded in the eighteenth century, however, when the new infinitesimal techniques developed by Newton (1665) and then independently by Leibniz (1675–6) were applied to the study of curves.

Initially, the invention of the calculus was understood as an extension of algebra to infinitely small quantities. This gave rise to the study of differential equations, so called because they contained not only variables, say x and y, but also their differentials dx, dy and even possible differentials of higher order, like ddx. Nevertheless, the mathematicians of this period mainly expressed the objects and methods of the calculus in geometric language. While they used integral calculus to compute areas under given curves (called quadratures), the related differential calculus gave them a procedure for finding tangent lines to curves, called the direct method of tangents. The inverse method of tangents meant finding a curve from given properties of its tangents, i.e., solving a certain type of differential equation (cf. insert 1). The integral calculus gave mathematicians the power to introduce hitherto unknown curves by the means of integral expressions obtained by solving such differential equations. These formal expressions represented quadratures of algebraic curves, but this led to a host of new problems, namely to find explicit solutions for these expressions. If they failed to obtain these, the practitioners of the time considered the corresponding solutions to represent non-algebraic or transcendental curves. This again opened up vast new horizons and had a huge impact on mathematics.

These developments were inextricably linked during the eighteenth century with concurrent interests in mechanics. Indeed, mathematicians striving to test the newly invented methods often applied them to problems from mechanics, by treating these in an idealized form that transformed the physical problems into mathematical entities and relationships. This method of formulating mechanical problems often led to differential equations, which required new analytical techniques to be integrated. These, in turn, enabled mathematicians to solve questions hitherto considered inaccessible in mechanics, and more generally in natural philosophy or physics. Curves and surfaces were fundamental to this idealized physics, as the calculus now made it possible to explore many of their properties (see Chapter 5).

The period under study in this volume opens with Newton's seminal *Philosophiae naturalis principia mathematica* (The Mathematical Principles of Natural Philosophy) published in 1687. In what follows, we focus on three main stages in the development of mathematics. The first concerns the expansion of the calculus at the turn of the century. Still deeply anchored in the problem-solving tradition of the seventeenth century, mathematicians tended to extend the methods of the newly created calculus by means of problem challenges, a frequent source of innovation. Following this stage of intense creativity came a period of stabilization, even if the polemics and prodding challenges among mathematicians did not disappear. Centered in the middle of the century, the activities during the second stage were characterized by a greater effort toward systematizing the considerable body of acquired knowledge. While attempting to put things in order, mathematicians recognized that their map of knowledge showed lacunas, a lack of foundations, and even contradictions, which needed to be treated. Mathematics was by then considered part of the enlightened culture that came to dominate large parts of the European continent. Finally, a subtle change will be described which took place in the second half of the century, from the investigation of single problems (often linked to mechanics or optics) to the study of more theoretical matters. The latter developments took the results already obtained as a reference point, without interfering necessarily with problems arising from the external world (e.g., mechanical problems). Instead, leading mathematicians tended to raise theoretical questions, asking, for example, about the existence of solutions of differential equations. They pondered why one equation has a unique solution while another does not. This third stage also saw the institutionalization of research mathematics not only in academies, but also in universities or other teaching institutions, which supported a more autonomous development of the discipline.

CHALLENGES AND DISPUTES EXPANDING THE CALCULUS

The calculus, as formulated by Newton and Leibniz, took two different forms, the Leibnizian form based on differentials (differences) and the Newtonian version, which employed the notion of fluxions (velocities). Curves were central to both approaches. Newton's understanding of curves was kinematic, that is to say that he understood curves as generated by motion. Points moved along a curve by a combined motion of its abscissae and ordinates (the x- and y-directions using Cartesian coordinates). The latter, called fluents, increased or decreased with point-wise or instantaneous speed, which Newton called fluxions. For him, curves (which he sometimes called crooked lines) had continuous curvature.

Leibniz instead placed emphasis on variables that range over a discrete sequence of incomparably near values; differentials are then the differences between contiguous pairs of such variables. In a famous paper, published in the Leipzig based *Acta eruditorum* (Scholars' Journal) in 1684, Leibniz introduced the efficient *d*-notation still in use today and presented the standard rules for differentiation. The Leibnizian calculus was an "analytical machinery for the study of curves" (Engelsman 1984: 4), which were in his eyes polygons with infinitely many edges. (For a more detailed presentation of the two methods, see Chapter 3 in Volume 3; Bertoloni Meli 1993; Bos 1974; and Guicciardini 2009.)

A Decade of Newtonian Publications

Mathematical invention in the eighteenth century was partly shaped by the open problems posed in Newton's groundbreaking *Principia*, written when he was Lucasian Professor of Mathematics at Cambridge. This classic was composed of three books, in which the first dealt with the motion of bodies in a void, the second with motion in a resisting medium, and the third culminated with the application of the mathematical results in Book 1 to celestial bodies subject to the principle of universal gravitation. For Newton, a devout Christian, this achievement led to something more profound than the invention of celestial mechanics; he understood this as a means to discover and understand how God had created the world (see Chapter 4). One might ask: what was mathematical in this treatise devoted to the description of the universe and the forces that act on the planets? For Newton, mathematics was a means to guarantee certainty in natural philosophy. As a fine and creative mathematician, he set forth in his *Principia* a vast array of methods for attacking a large repertoire of open problems. It is difficult to render full justice to Newton's multifaceted and at times somewhat incoherent approach to mathematics. That he was "a great problem solver" (Guicciardini 2009: xvii) can easily be understood by anyone who wishes to page through the eight volumes of his *Mathematical Papers* (edited by Derek T. Whiteside, 1967–81). It should be noted that a great deal of the material found in those volumes had never seen the light of day until this modern edition appeared.

During his lifetime, Newton had been most reluctant to publish his discoveries for a variety of reasons, as several twentieth-century scholars have described. A major factor in this context was his philosophical agenda. Newton considered his method of fluxions to be a heuristic tool, useful in the art of discovery, but not standing up to the high status he assigned to mathematics. Faithful to this idea, he shared his early inventions by circulating some of his manuscripts within a very small circle of British mathematicians. In his printed publications, addressing a wider reading public, he resorted to demonstrations using geometry, whose certainty and elegance he revered. In the *Principia*, he only briefly mentioned what he called his "new analysis," but otherwise he avoided the use of infinitely small quantities, which were not properly defined and thus controversial. Instead, he opted for a kind of geometric limit procedure, the so-called method of first and ultimate ratios of vanishing quantities, set forth at the beginning of Book 1. This enabled him to prove, for example, that the arc-lengths, tangents, and chords will ultimately become equal as they "vanish" or continuously decrease to zero. As a consequence, those who had the mathematical ability needed to read Newton's *Principia* had no chance of learning the first principles of the calculus from it.

After the turn of the century—by which time Newton was no longer productive as a mathematician, having already left Cambridge in 1696 to become Warden of the Mint in London—two of his earlier works appeared in 1704 as appendices to his *Opticks*. Newton's tract entitled *Enumeration of Lines of the Third Order* gave the first classification of cubic curves, whereas *On the Quadrature of Curves*, first drafted in 1691–2, provided the first public presentation of some of his earlier work on integral calculus. In a statement about the nature of mathematical objects, Newton rejected infinitely small figures and introduced the dotted notation for fluxions, ($\dot{x}$), which express the rate of change of a continuous variable, x, usually with respect to time. This small treatise was devoted to finding the quadratures of families of algebraic curves, either in terms of finite algebraic equations or by reducing the quadrature sought to others already known or familiar cases that were unknown, such as the quadratures of the circle and the hyperbola. Newton presented the results of his quadratures in ready-to-use tables. Thus, his focus in this treatise was not on a systematic presentation of his method of fluxions, nor did he deal with the direct problem of finding the fluxions when the fluents were given (i.e., differentiation in the Leibnizian sense). His aim rather was restricted to the inverse problem of finding the fluents when the fluxions are given (integration or finding antiderivatives), a problem he considered—and rightly so—more difficult than finding derivatives. Moreover he considered that the inverse problem was not completely settled and still an open question.

Another mathematical tract, *De analysi* (On Analysis), although written much earlier, first appeared in print in 1711 (see Figure 3.1) on the brink of the explosive calculus debates. William Jones published Newton's text in his *Analysis per quantitatum series, fluxiones, ac differentias* (Analysis for Quantitative Series, Fluxions and Differences), before it was reprinted in the infamous *Commercium epistolicum* (1712), discussed below. In this tract, Newton mainly introduced his method of series[1] of which he was very proud. This enabled him to expand y—whether expressed by a fraction $\frac{p(x)}{q(x)}$, a root $\sqrt{p(x)}$ or implicitly by an algebraic equation $f(x,y) = 0$—as a power series, treated like a polynomial with a possibly infinite sum of terms of the form ax^n. At least since 1635, mathematicians knew how to calculate the areas under curves of the type $y = ax^n$. Assuming that the sum of areas is equal to the area of the sum, this knowledge provides a general method enabling the quadrature of a large class of curves, and for Newton this was the core idea behind his innovation.

These two printed treatises provided a biased and highly incomplete vista of Newton's mathematical achievements (Malet 2020: 17). Nevertheless, it was mainly through them that Continental European mathematicians were made aware of the fact that Newton had invented a method, which served purposes similar to methods based on Leibnizian differentials. Newton's most explicit and complete treatise, entitled *On the Method of Fluxions and Infinite Series*, was written in Latin in 1671, presumably for Newton's personal use. It only appeared posthumously, however, in 1736 in an English translation, followed in 1740 by a French translation (from the English), prepared by Georges-Louis Buffon. Had this masterpiece already appeared in the 1670s, it very likely would have revolutionized mathematics, but more than half a century later it contained nothing not already well known and established by Leibnizian methods. This single example nicely illustrates how the circulation of

FIGURE 3.1. Engraving from William Jones, ed. Isaac Newton, *Analysis per quantitatum series, fluxiones, ac differentias: cum enumeratione linearum tertii ordinis* (London: Pearson, 1711, p. 1), illustration of the title page of *De analysi*. The diagrams represented on the scrolls and shields are taken from Newton's *Principia*. For a detailed interpretation of this picture, cf. Guicciardini (2009: 370). Photo courtesy of Library of Congress, Rare Book and Special Collections Division.

mathematical knowledge was an integral part of a complex production process. The channels of communications in this era were few and they operated very slowly, often leading to loss of important results. A result, however important, if kept secret or circulated only locally, would then leave no trace or notable effect on the development of the discipline.

Extending Leibnizian Calculus

While only a few knew about the Newtonian method of fluxions, by the end of the seventeenth century, the Leibnizian calculus had already made a major impact. Soon several leading European geometers would master its basic techniques, while extending and refining them in the process. Among this group, the Swiss Bernoulli brothers, Jacob and Johann, and the Parisian Marquis de l'Hôpital were particularly prominent. Jacob Bernoulli, trained as a theologian and since 1687 professor of mathematics at the University of Basel, introduced his much younger brother, Johann, then a student in medicine at the same university, to mathematics. Their works were closely intermingled and strongly marked by a growing competition and rivalry between them. The Bernoulli brothers were among the first, along with the Scot John Craig, to use the Leibnizian notation and algorithms. Over the years, the Bernoullis acquired considerable analytical skills, especially with quadratures, and both made important contributions to mathematics. In his first paper devoted to the calculus, published in May 1690 in the *Acta eruditorum*, Jacob applied the calculus to the problem of determining the Leibnizian paracentric isochrone: i.e., the curve along which an object will descend uniformly and approach the horizon in such a way that equal heights are traversed in equal times (i.e., with a constant vertical velocity). Bernoulli showed that the problem was equivalent to solving a differential equation (see Insert 1) and was able to find the solution in a closed form $y = y(x)$, which he identified to be the equation of a semi-cubic parabola. In the solution process, he managed to separate the variables (i.e., to rewrite the equation in the form $dx = f(y)dy$, a technique recently invented by Leibniz). Still clinging to geometrical representations like most of his contemporaries, he interpreted the equation as an equality of two (infinitesimal) rectangles, $1 \cdot dx$ and $f(y) \cdot dy$, hence "their integrals will also be made equal." This was the very first occurrence of the term "integral"; Leibniz spoke of sums (of differences) and Newton of areas or quadratures.

Insert 1: Differential Equations

A differential equation is an equation which contains not only variables x, y but also their differentials, or differences in Leibnizian language, dx, dy, and even differentials of a higher order, say ddx. A very simple example would be $\frac{dy}{dx} = 0$. To solve a differential equation is to discover the so-called integral curve y. In our example, the integration of the differential equation is straightforward: the solution is $y(x) = C$ = constant, (as $dy = dC = 0$), i.e., an infinity (or family) of straight horizontal lines. In the geometrical language of the early eighteenth century, this means that the solution curves all have the property that their slope (or tangent) is zero. To obtain a uniquely determined solution, a single horizontal line, we need an additional condition, like $y(5) = 64$, or $y(x_0) = y_0$. To solve a differential equation turned out to be more complicated than simple integration, as the integral curve cannot always be found in closed form $y = y(x)$, and demanded from eighteenth century mathematicians high analytical skills.

This kind of problem solving was emblematic for the early Leibnizian calculus. Other examples concerned the shape of a rope fixed at two points in the same plane (catenary) or the shape of a sail under the pressure of wind, a problem whose mathematical formulation involved the solution of a second-order differential equation (see below). These types of problems were extremely difficult to solve without the calculus. Alongside Leibniz, with whom they were in close epistolary contact, the Bernoulli brothers excelled in developing skills and tools to solve differential equations, as they found that many interesting physical problems could be expressed in terms of these. After an initial period of collaboration, a growing sense of competition arose between the two

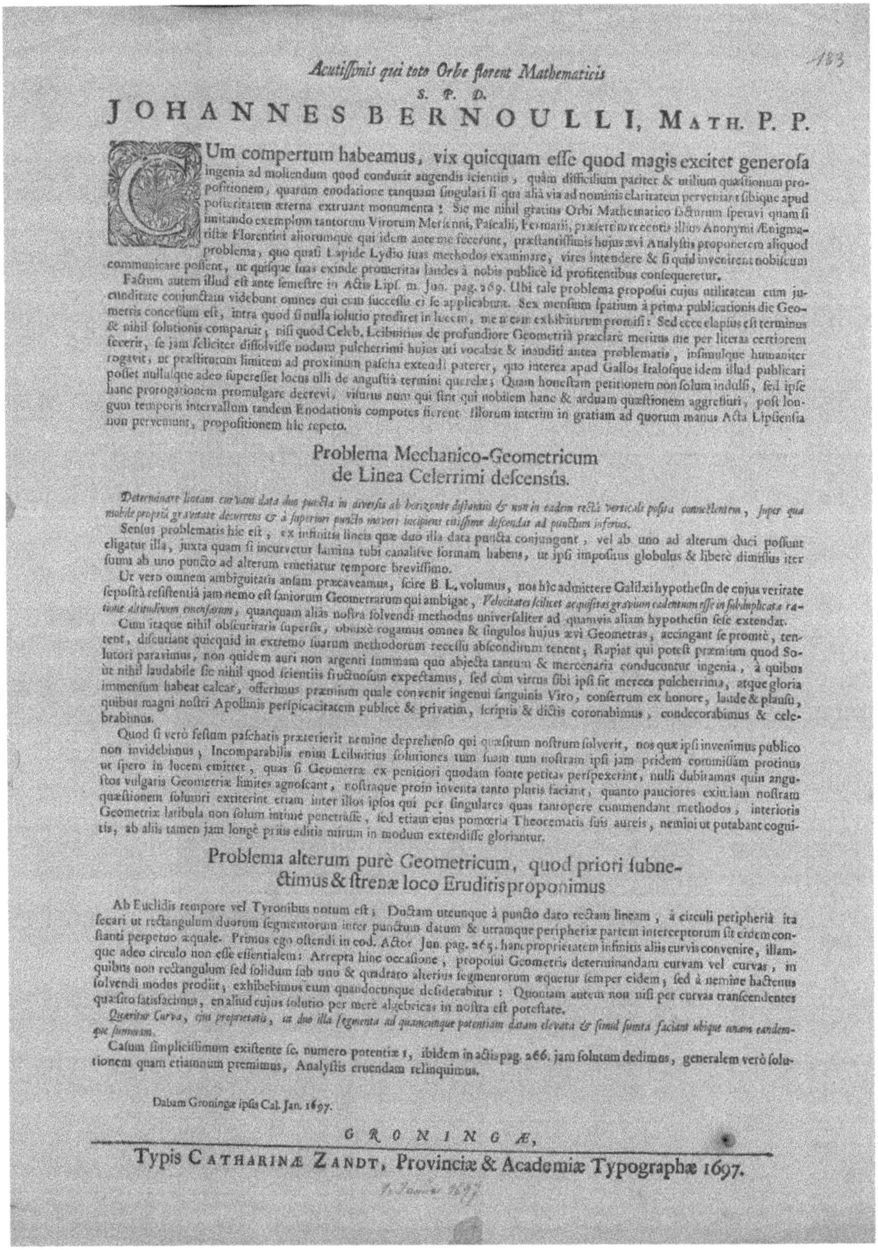

Acutissimis qui toto Orbe florent Mathematicis

S. P. D.

JOHANNES BERNOULLI, MATH. P. P.

CUm compertum habeamus, vix quicquam esse quod magis excitet generosa ingenia ad moliendum quod conducit augendis scientiis, quàm difficilium pariter & utilium quæstionum propositionem, quarum enodatione tanquam singulari si qua aliâ via ad nominis claritatem perveniant sibique apud posteritatem æterna extruant monumenta; Sic me nihil gratius Orbi Mathematico facturum speravi quam si imitando exemplum tantorum Virorum Mersenni, Pascalii, Fermatii, præsertim recentis illius Anonymi Ænigmatistæ Florentini aliorumque qui idem ante me fecerunt, præstantissimis hujus ævi Analystis proponerem aliquod problema, quo quasi Lapide Lydio suas methodos examinare, vires intendere & si quid invenirent nobiscum communicare possent, ut quisque suas exinde promeritas laudes à nobis publicè id profitentibus consequeretur.

Factum autem illud est ante semestre in Actis Lips. m. Jun. pag. 269. Ubi tale problema proposui cujus utilitatem cum jucunditate conjunctam videbunt omnes qui cum successu ei se applicabunt. Sex mensium spatium à prima publicationis die Geometris concessum est, intra quod si nulla solutio prodiret in lucem, me eam exhibiturum promisi: Sed ecce elapsus est terminus & nihil solutionis comparuit, nisi quod Celeb. Leibnitius de profundiore Geometriâ præclarè meritus me per literas certiorem fecerit, se jam feliciter dissolvisse nodum pulcherrimi hujus uti vocabat & inauditi antea problematis, insimulque humaniter rogavit, ut præstitutum limitem ad proximum pascha extendi paterer, quo interea apud Gallos Italosque idem illud publicari posset nullusque adeo superesset locus ulli de angustiâ termini querelæ; Quam honestam petitionem non solum indulsi, sed ipse hanc prorogationem promulgare decrevi, visurus num qui sint qui nobilem hanc & arduam quæstionem aggressuri, post longum temporis intervallum tandem Enodationis compotes fierent. Illorum interim in gratiam ad quorum manus Acta Lipsiensia non perveniunt, propositionem hic repeto.

Problema Mechanico-Geometricum de Linea Celerrimi descensûs.

Determinare lineam curvam data duo puncta in diversis ab horizonte distantiis & non in eadem rectâ verticali posita connectentem, super qua mobile propriâ gravitate decurrens & à superiori puncto moveri incipiens citissime descendat ad punctum inferius.

Sensus problematis hic est, ex infinitis lineis quæ duo illa data puncta conjungunt, vel ab uno ad alterum duci possunt eligatur illa, juxta quam si incurvetur lamina tubi canalisve formam habens, ut ipsi impositus globulus & liberè dimissus iter suum ab uno puncto ad alterum emetiatur tempore brevissimo.

Ut vero omnem ambiguitatis ansam præcaveamus, scire B. L. volumus, nos hîc admittere Galilæi hypothesin de cujus veritate sepositâ resistentiâ jam nemo est saniorum Geometrarum qui ambigat, *Velocitates scilicet acquisitas gravium cadentium esse in subduplicata ratione altitudinum emensarum*, quanquam aliàs nostra solvendi methodus universaliter ad quamvis aliam hypothesin sese extendat.

Cum itaque nihil obscuritatis supersit, obnixè rogamus omnes & singulos hujus ævi Geometras, accingant se promtè, tentent, discutiant quicquid in extremo suarum methodorum recessu absconditum tenent; Rapiat qui potest præmium quod Solutori paravimus, non quidem auri non argenti summam quo abjecta tantum & mercenaria conducuntur ingenia, à quibus ùt nihil laudabile sic nihil quod scientiis fructuosum expectamus, sed cùm virtus sibi ipsi sit merces pulcherrima, atque gloria immensum habeat calcar, offerimus præmium quale convenit ingenui sanguinis Viro, consertum ex honore, laude & plausu, quibus magni nostri Apollinis perspicacitatem publicè & privatim, scriptis & dictis coronabimus, condecorabimus & celebrabimus.

Quod si verò festum paschatis præterierit nemine deprehenso qui quæsitum nostrum solverit, nos quæ ipsi invenimus publico non invidebimus; Incomparabilis enim Leibnitius solutiones tum suam tum nostram ipsi jam pridem commissam protinus ut spero in lucem emittet, quas si Geometræ ex penitiori quodam fonte petitas perspexerint, nulli dubitamus quin angustos vulgaris Geometriæ limites agnoscant, nostraque proin inventa tanto pluris faciant, quanto pauciores existiam nostram quæstionem soluturi extiterint etiam inter illos ipsos qui per singulares quas tantopere commendant methodos, interioris Geometriæ latibula non solum intimè penetrasse, sed etiam ejus pomœria Theorematis suis aureis, nemini ut putabant cognitis, ab aliis tamen jam longè prius editis mirum in modum extendisse gloriantur.

Problema alterum purè Geometricum, quod priori subnectimus & strenæ loco Eruditis proponimus

Ab Euclidis tempore vel Tyronibus notum est; Ductam utcunque à puncto dato rectam lineam, à circuli peripheriâ ita secari ut rectangulum duorum segmentorum inter punctum datum & utramque peripheriæ partem interceptorum sit eidem constanti perpetuo æquale. Primus ego ostendi in eod. Actor. Jun. pag. 265. hanc proprietatem infinitis aliis curvis convenire, illamque adeo circulo non esse essentialem: Arrepta hinc occasione, proposui Geometris determinandam curvam vel curvas, in quibus non rectangulum sed solidum sub uno & quadrato alterius segmentorum æquetur semper eidem; sed à nemine hactenus solvendi modus prodiit; exhibebimus eum quandocunque desiderabitur: Quoniam autem non nisi per curvas transcendentes quæsito satisfacimus, en aliud cujus solutio per merè algebricas in nostra est potestate.

Quæritur Curva, ejus proprietatis, ut duo illa segmenta ad quamcunque potentiam datam elevata & simul sumta faciant ubique unam eandemque summam.

Casum simplicissimum existente sc. numero potentiæ 1, ibidem in actis pag. 266. jam solutum dedimus, generalem verò solutionem quam etiamnum premimus, Analystis eruendam relinquimus.

Dabam Groningæ ipsis Cal. Jan. 1697.

GRONINGÆ,

Typis CATHARINÆ ZANDT, Provinciæ & Academiæ Typographæ 1697.

FIGURE 3.2. Johann Bernoulli's placard advertising the challenge problem (brachistochrone), printed in Groningen, January 1697. Photo courtesy of Universitätsbibliothek Basel (UBH KiAr GI7:X:183).

brothers, who proposed more and more difficult challenge problems to each other, some of which germinated into important mathematical developments.

One such challenge problem asked to find the curve of swiftest descent (or brachistochrone), today highly appreciated by young skateboarders. Victor Katz called this "the most significant of the problems proposed by Johann Bernoulli, in terms of its ultimate consequences for mathematics" (Katz 2009: 586). Johann was well aware of its significance and circulated the problem widely, in letters, learned journals and even as a placard (see Figure 3.2).

In 1696, Johann Bernoulli, who was then professor of mathematics in Groningen (The Netherlands), formulated this challenge in an issue of the *Acta eruditorum*. There he wrote, "If two points A and B are given in a vertical plane, to assign to a mobile particle *M* the path *AMB* along which, descending under its own weight, it passes from the point *A* to the point *B* in the shortest time." Although he formulated this as a simple-sounding minimum problem, something mathematicians had been handling since the seventeenth century, finding the brachistochrone curve meant solving a problem of a quite different nature. It involved finding the unique curve satisfying this minimum property among an infinite number of other curves that had the same extremities, A and B, but were otherwise unknown. Thus, the problem led to an investigation of families of curves, which Leibniz had already begun to study. This, in turn, would eventually lead to fundamental new concepts in the theory of functions of several variables.

The curve of swiftest descent, having its origin in A and passing through B, turned out to be an inverted cycloid. The cycloid is a curve generated by a point fixed on a circumference of a circle that rolls along a straight line, very well known to geometers of the time. Johann's solution relied on a clever analogy with optics. Fermat showed that Snell's Law for refraction follows from a minimum principle, namely the assumption that the path traversed by a light ray passing from a thinner to a denser medium will take the least time among all possible paths. In applying it for the case of the brachistochrone problem, Bernoulli ingeniously assumed that the vertical plane of the problem was composed of infinitesimally thin horizontal layers whose densities varied continuously. The brachistochrone, the path of least time, was then analogous to a light ray emitted from A whose direction changed continually as it passed from one layer through to the next. At every point the sine of the angle between the tangent to the curve and the vertical axis was proportional to the velocity (Snell's law), and the velocity was in turn proportional to the square root of the distance fallen (Galileo's free fall law). This mathematical relationship yielded a differential equation, which Johann immediately recognized as corresponding to cycloid curves.

Johann's solution produced other results concerning families of curves as well. He considered all brachistochrone cycloids that originate in A but have different endpoints B and imagined the curves generated by releasing different particles from A at the same time (see Figure 3.3). At each instant of time, these

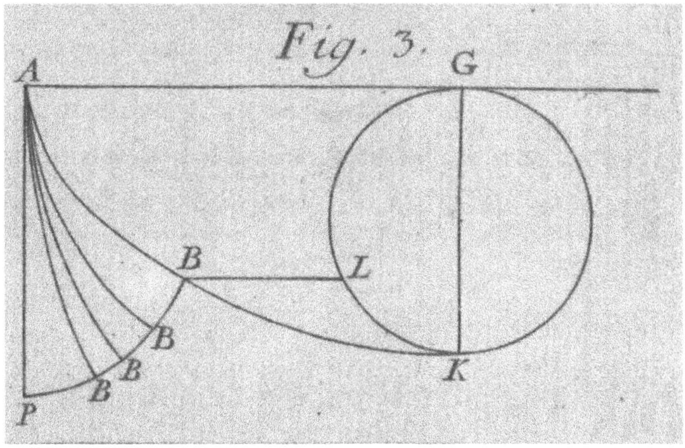

FIGURE 3.3. Figure from Bernoulli 1697, a cycloid is developed (ABK), the trajectories orthogonal to the cycloid are isochrones (PB). Photo courtesy of ETH-Bibliothek Zürich, Rar 5026: 1. Public Domain.

particles lie on a curve (PB), and the family of all such curves are called synchrones. Turning back to his optical model, these synchrones correspond to curves formed by the simultaneous positions of light pulses originally emitted from A; hence, these are wave fronts. According to Christiaan Huygens' theory of light, wave fronts were always perpendicular to the direction of propagation of light rays, and this meant that the synchrones (PB) were the orthogonal trajectories of the family of cycloids (AB).

This type of problem—finding the orthogonal trajectories to a given family of transcendental curves—soon led to the development of ideas about partial differentiation (see below). Moreover, this famous example illustrates a characteristic feature of early eighteenth-century mathematics, namely the fruitful interplay of resources, here combining mechanics, optics and geometry. Jacob's solution of the brachistochrone problem arose from a different way of thinking about how to solve this specific problem. More cumbersome than Johann's elegant proof, but also more easily generalizable, Jacob's solution contained the germs of a general method that had a long afterlife. It was to develop into a new branch of mathematics, the calculus of variations, established by Euler in 1744 along the lines put forward by Jacob Bernoulli, before Joseph-Louis Lagrange switched to a more analytical approach (also discussed below).

A Shared Mathematical Culture

The turn of the century brought about a period of rather strong competition among European mathematicians, whether from rival nations or feuding family members, like the Bernoulli brothers. The practice of challenging competitors accorded well with the publishing style chosen by the early analysts, who preferred writing short articles, many of which circulated in newly created scholarly journals. The most important vehicle for Leibnizian analysis was the *Acta eruditorum*, founded in 1682. On the Continent, a small community of its practitioners emerged, sharing the same conceptual framework, language and set of problems, which they discussed in their correspondence and published in Latin in that same journal.

This common mathematical culture also benefitted from the early publication of a highly influential textbook in differential calculus, the *Analyse des infiniment petits pour l'intelligence des lignes courbes* (Infinitesimal Analysis for the Comprehension of Curved Lines), composed and then published anonymously in 1696 by the French nobleman, Guillaume François Antoine, Marquis de l'Hôpital. Following a brief military career, L'Hôpital had devoted himself to studying mathematics. As it happened, the young Johann Bernoulli was visiting Paris at the beginning of the 1690s on his *Grand Tour* of studies, and he then offered to teach the Marquis de l'Hôpital the key ideas of the Leibnizian calculus. Later, it came to light that the material presented in the *Analyse*, as well as the structure of the book, were to a very large extent based on the private lessons Bernoulli gave to its author. In the introduction, L'Hôpital admitted to having made free use of the discoveries of Leibniz and the Bernoulli brothers, but without acknowledging his full debt to Johann.

At the outset, L'Hôpital introduced variable quantities as those that continually increase or decrease, and he defined a differential as the infinitely small part by which a variable quantity increases or decreases. L'Hôpital apparently never questioned the existence of infinitesimals, although both Newton and Leibniz tried to avoid these in their writings, as they were still a thorny subject at that time. He represented infinitesimals geometrically and asserted that they satisfied the rules given by Leibniz in his 1684 paper. He then showed how they enable one to study the properties of curves, such as tangents, maxima and minima, points of inflection, cusps, radii of curvature, evolutes, etc. L'Hôpital's treatment was mainly restricted to algebraic curves, except for mention of the logarithmic curve. His presentation was highly geometrical; most quantities, finite or infinite, even second-order differentials, appeared in carefully drawn diagrams. The book was entirely devoted to the differential calculus without any reference to methods for treating problems in the inverse integral calculus. L'Hôpital's *Analyse* circulated widely, not only in France but also in the British Isles. Eight years later, Charles Hayes published his *Treatise on Fluxions* (1704), a calculus text dedicated to the Director of the Royal African Company of England. In it, he borrowed freely from all available mathematical sources, including L'Hôpital's text.

Leibniz and the Bernoulli brothers in Basel were highly successful in spreading the new infinitesimal calculus. The mid-century Uppsala professor

Samuel Klingenstierna was instrumental in introducing it to Sweden. Although he was aware of Newton's method of fluxions, Klingenstierna's mathematical works were strongly influenced by Johann Bernoulli and consequently adopted Leibnizian differential calculus. In Italy, Jacob Hermann a student of Jacob Bernoulli, then Nicolaus Bernoulli, Jacob's and Johann's nephew, were successively appointed with the strong support of Leibniz to the chair of mathematics at the University of Padua where they promoted the new analysis. Their teaching eventually called into existence a circle of fine practitioners of Leibnizian calculus in Italy, including Jacopo Riccati and Giovanni Poleni, the Tuscan Guido Grandi and Gabriele Manfredi in Bologna. In the Republic of the Netherlands, Newton's name stood higher than that of Leibniz, though mainly for religious reasons. The Dutch revered his *Principia* as the key that unlocked the mysteries of God's creation. The divide separating the Newtonians on the British Isles and the Leibnizians on the Continent was less clear-cut than traditionally admitted. What gradually circulated throughout much of Europe was a blend of Newtonian and Leibnizian methods. Mathematicians picked up notions and techniques from both traditions, many following their own research agendas and cultural contexts (see Pulte and Mandelbrote 2019 for the reception of Newton and Newtonianism). The circulation of Newtonian and Leibnizian ideas outside Europe, on the other hand, has received little attention and remains a relatively open problem in the history of mathematics.

One consequence of the successful spread of Leibnizian analysis on the European Continent was in helping to produce a competent reading public for Newton's *Principia*. Since most Continental mathematicians at the turn of the century grew to maturity in the problem solving tradition, they found many enticing things in the *Principia*, especially when it came to problems dealing with orbital motions of heavenly bodies. The Paris Academician Pierre Varignon, correspondent of Leibniz and the Bernoullis, wrote a series of papers (1700–6) devoted to central forces, either centripetal (attraction toward a central point) or centrifugal forces, which Huygens had studied earlier. In the *Principia*, Newton conceived of planetary motion in the void as resulting from gravitational central forces, where gravity acted at a distance and instantaneously.

Varignon was primarily interested in the mathematization of the forces that produced these orbital motions, which involved taking into account the curvature of the trajectories of the planets expressed in terms of second-order differentials. Different mathematical representations of trajectories were then in use. In the *Principia*, these were continuous and represented locally by circular or parabolic arcs representing accelerated motion. Continentals, by contrast, treated trajectories as polygons with infinitesimal sides, along which planets moved uniformly while receiving instantaneous impulses at each of the vertices. Varignon understood that both representations corresponded to correct mathematical models, but that mixing them could also lead to errors.

Furthermore, he found that the force that caused a body to revolve around the focus of a conic satisfied Newton's gravitational law (or inverse-square law). Although this result was not new—it was equivalent to Propositions 11–13 in Book 1 of the *Principia*—Varignon formulated this in terms of Leibnizian differentials. This process of translating the geometrical language of the *Principia* into the idiom of Leibnizian differentials proved highly fertile, in so far as it allowed the Continentals to make considerable advances in the application of mathematics to dynamics (see Chapter 5).

The Priority Dispute as Battleground

Precisely because the Newtonian and Leibnizian formulations of the calculus were used to perform analogous operations and to solve the same range of problems—essentially drawing tangents, finding extreme values, calculating curvature, obtaining quadratures, computing the length of curves (rectifications)—the question of priority could be raised and opened a dramatic dispute, during which the principals along with their respective supporters violently clashed. Already in the 1690s, Wallis complained about the fact that the "notions of fluxions" were circulating on the Continent "by the name of Leibniz's *calculus differentialis.*" Johann Bernoulli expressed the opposite complaint after having read some fragments of Newton's method in Wallis' *Opera* (1693). Later, the character of the accusations turned much more openly hostile and personal. First came a paper in 1699 on the brachistochrone problem, written by the Swiss mathematician Nicolas Fatio de Duillier, then in 1708, John Keill published a paper on central forces in the *Philosophical Transactions of the Royal Society of London.* In both cases, the authors accused Leibniz publicly of having plagiarized Newton. Leibniz, as a foreign member of the Royal Society, took for granted that Newton, its president, would never support such a claim. He thus requested that the Royal Society officially discharge him from the more recent accusation.

The response was the infamous *Commercium epistolicum* (1712), in which Newton himself was deeply involved in support of Keill's accusation. He collected evidence for it from the textual records held in the archives of the Society, including letters exchanged with Leibniz during the mid-1670s; he also drew material from the extensive correspondence of John Collins, who had been Leibniz's principal mathematical contact during the time of his visits to London. The *Commercium* was a dazzling montage of excerpts from various documents, woven together in a narrative that aimed to prove "Newton was the first inventor of the calculus and that Leibniz's differential method is one and the same with the method of fluxions, excepting the name and mode of notation" (*Commercium epistolicum* 1712: 121). Exegetical footnotes guided the reader and offered a key to a partisan interpretation, which deprived Leibniz of any right to the title of co-inventor of the calculus. The heated controversy between Newton and Leibniz that followed publication of the *Commercium* degenerated into a fierce confrontation between the British and the Continentals, punctuated by mathematical challenges, personal rivalries and harsh attacks with nationalist overtones. We cannot enter into the details of the numerous episodes of this bitter quarrel, and refer the reader to the literature (Hall 1980; Guicciardini 2018b; Sonar 2018). Instead, we here briefly describe the mathematical issues that were then at stake.

First, there was no consensus regarding the nature of the calculus. In fact, both parties disagreed over what each exactly claimed to have discovered. During the dispute, Newton placed his discoveries on infinite series to the fore, including their use for solving quadrature problems, whereas Leibniz understood the calculus as an algorithm based on an efficient notation for the differential calculus. He evaluated quadratures as sums of area differentials and was well aware that no general algorithm existed for the integral calculus. As already noted, this led to the search for various specific techniques. Johann Bernoulli was highly skilled in this "craft" (Guicciardini, 2018b: 11), trusting this would enable him to win the numerous contests he launched during the twenty years that this struggle lasted.

Second, Leibnizians and Newtonians had divergent understandings on what it meant to possess a valid solution of a problem. The discussions concerning central forces, over which Keill and Johann Bernoulli argued in the second decade of the century, provide a good example. In his *Principia* (I.41), Newton had "solved" the problem of finding the trajectory of an orbiting body when the law of the centripetal force is given, "granted the quadratures of curvilinear figures." This meant that Newton had "reduced the problem to quadratures," which he then assumed to be within reach. He thus took for granted that his methods led to a solution of the equation for the trajectory, though he did not provide it in the *Principia*. Bernoulli, who studied the calculation of planetary orbits intensely, had shown that if the force is inversely proportional to the square of the distance, the only possible trajectories are conic sections. He accomplished this by obtaining a differential equation for the orbits, which he then integrated to exhibit the conic traversed by the planet. Bernoulli argued that Newton had not "solved" the problem, whereas Newton's champion, Keill, thought that Bernoulli had merely translated Newton's solution into a more algebraic idiom.

Third, underlying these debates were fundamentally different views about what made the value of the calculus. Leibniz was fully aware of the efficiency of a convenient notation and algorithmic rules that one could follow blindly. Indeed, he achieved to such a system. Vincenzo Viviani, a stout supporter of classical mathematics, once ironically compared this to a children's game, but played with symbols rather than with nuts. Newton, by contrast, attached little value to symbols in mathematics. He considered geometry as the only acceptable language when it came to publishing rigorous and elegant proofs. This conviction, however, did not prevent him using algebra as a heuristic tool in his writings.

Among the significant mathematical results obtained during the priority dispute, we can mention the brilliant solution proposed by Johann Bernoulli for the motion of a projectile in a medium whose resistance is proportional to any power of the velocity. He also solved the general problem of finding orthogonal trajectories to a family of curves. However, one of the most important outcomes concerned the development of new techniques for the integration of differential equations. Among the Newtonians, creative mathematicians, like Abraham de Moivre, Roger Cotes, and Brook Taylor, realized that integration was an alluring and open field of research. In his *Methodus incrementorum directa & inversa* (Direct and Inverse Method of Increments, 1715), Taylor interpreted the calculus as a limiting case of a theory of finite increments. He applied this method to the solution of mechanical problems, most of them familiar and already solved. His book also presented the famous Taylor expansions, power series involving higher derivatives evaluated at a given point, which the Scot Colin Maclaurin in his influential *Treatise of Fluxions* (1742) took to be the origin.

Having described the expansion of the calculus during the first decades of the century, we should also mention that from early on its soundness came under strong attacks. The logical difficulties connected with the infinitely small, but also the method of simply neglecting differentials of higher order without any formal justification, presented difficult issues that led to numerous debates right up until the early nineteenth century (Jahnke 2003: 127; and Chapter 5).

MID-CENTURY STABILIZATION

Encyclopedic Collections

Even if contests among mathematicians did not disappear along with the first generation of Newton's acolytes and Leibniz's disciples, the middle of the century saw a tendency towards collecting and organizing mathematical results in a systematic way. This fit into a context, in which many flourishing intellectual ventures sought to organize and systematize the expanding knowledge base. By mid-century, one grandiose enterprise overshadowed all others, namely the famous *Encyclopédie*, led by Denis Diderot and Jean d'Alembert and published between 1751 and 1772. To a considerable extent, its fame derives from the role it played in a much larger phenomenon, as it reflected the concerns of Enlightenment culture with the communication of knowledge. Indeed, a governing principle of the Enlightenment called for a circulation of information and ideas, among not only scholars but also extending to a larger audience.

Mathematics was considered as an integral part of the ideal of encyclopedic knowledge, just like painting or music. Interested readers could find mathematical knowledge in learned journals, dictionaries and encyclopedic collections across Europe and beyond. One instance of this phenomenon in a quite different context points to China, where the compilation of a collection of classic texts took place during the eighteenth century under the patronage of the Qing emperor Qianlong. This led to the Complete Library of the Four Branches of Books (*Siku quanshu*), in which mathematics appeared under the heading "philosophy." This project, under the editorial supervision of Dai Zhen, provided ample material for Chinese savants, whose principal preoccupation over the next century would be the interpretation of the ancient texts collected by Dai.

Encyclopedias often turned backwards to collect past knowledge, but they could also point to the future by providing information about significant advances. For mathematics, this proved to be the case with the *Encyclopédie*, the most prominent reference work of the French Enlightenment. Jean le Rond d'Alembert, then better known as a philosopher than as a mathematician, took charge of the entries on mathematics. As the natural-born son of a Parisian *salonnière*, d'Alembert was raised by the wife of a glazier, whereas his surrogate father took care that he received a solid education. He went on to study law and medicine, although his real passion was for mathematics. In 1739, d'Alembert presented his first paper to the Paris Academy of Sciences, to which he gained membership two years later, a position of central importance throughout his life. Although he published important contributions to dynamics, fluid mechanics and other parts of applied mathematics, his approach was purely mathematical.

D'Alembert's "Preliminary Discourse" for the *Encyclopédie* brought him great fame, though he contributed around 1,700 entries altogether, 1,500 of which are scientific in nature. These articles reported on new results and research programs, often in the form of doubts or objections, while drawing on the works of others. Thus, several entries stem from Émilie du Châtelet's *Institutions de physique* (Foundations of Physics) (Maglo 2019: 805). Incorporating entire passages taken from the recent mathematical literature into reference works was hardly unusual during this era. For example, John Harris, who belonged to Newton's circle, inserted an English translation of the latter's *On Quadrature* in his *Lexicon technicum* (Technical Lexicon, London, 1710). Christian Wolff followed the same practice in his *Vollständiges Mathematisches Lexicon* (Complete Mathematical Lexicon, Leipzig, 1716).

Yet, some of d'Alembert's entries in the *Encyclopédie* represented remarkable steps forward. A number of his longer contributions—articles on "Irreducible case," "Equation," etc.—had they been published together in a judicious non-alphabetical order would have formed an original treatise on algebra (which he never wrote in book form). D'Alembert also used the opportunity to clarify mathematical notions and took part in debates on the foundations of the calculus. The English mathematician and engineer Benjamin Robins had tried to clarify Newton's somewhat obscure expression "first and last ratios" by introducing the notion of limit as the quantity towards which a ratio of varying quantities tends, without ever reaching it. D'Alembert, in the entry "Differential" (*Encyclopédie*, vol. IV, 1754) defined the differential ratio dy/dx as the limit of the ratios of finite increments of y and x when these increments tend towards 0. His notion of limit, though not formally defined, was intuitively clear:

> a quantity is said the limit of another one, when the second can come closer to the first than any given quantity, however small it may be assumed, without the approaching quantity ever being able to surpass the quantity it is approaching; hence the difference of such a quantity from its limit is absolutely unassignable. (Entry "Limit," Encyclopédie, vol. IX, 542)

Calculus Treatises

Textbooks and treatises offered another form of more systematic presentation of mathematical results. As already mentioned, L'Hôpital's influential 1696 textbook was devoted solely to differential calculus, whereas integral calculus was included in a book written by the Milanese mathematician Maria Gaetana Agnesi, *Instituzioni analitiche ad uso della gioventù italiana* (Methods of Analysis for the Use of Italian Youth, 1748). Agnesi had studied the major mathematical works of her time, often scattered over various books, academic proceedings and journals. She also discussed these works in epistolary exchanges with a number of mathematicians, but especially with the Riccati brothers, and compiled this information in personal notes that served as a basis for her book. Addressed to Italian youth, it clearly explained the main notions of the Leibnizian calculus, while providing numerous examples. This work enjoyed great success in Italy as well as abroad, after translations appeared in both French (1775) and English (1801).

While Agnesi's *Methods* came at the end of the developments that took place during the first half of the century, Euler's *Introductio in analysin infinitorum* (Introduction to Analysis of the Infinite, 2 vols), also published in 1748, represented the starting point for a completely new approach to analysis. Euler's *Introduction* provided the foundational knowledge required to study his version of the calculus (what some call algebraic analysis). In his view, the calculus should be founded on the notion of function. That's why be gave in his treatise the elements of a theory of functions. This view was to thoroughly reshape what Leibniz had considered a new mathematical branch, namely analysis.

Leonhard Euler (see Figure 3.4) was one of the major figures of eighteenth-century mathematics. Trained in Basel by Johann Bernoulli, in 1727 at the age of not even twenty he joined the Academy of Sciences in Saint Petersburg founded three years earlier by Tsar Peter the Great as part of his efforts to modernize Russia. In St. Petersburg, Euler benefitted from an excellent research environment at the Academy during the early years of the institution. As a highly prolific writer endowed with a prodigious memory, he contributed significant advances to a large range of scientific disciplines, including analysis, calculus of variations, algebra, number theory, mechanics, and hydrodynamics. His recent biographer, Ronald S. Calinger (2016: 203) has described after André Weil Euler's approach to mathematics as experimental; he made exhaustive calculations until reaching a point when his methods no longer enabled him to go further. Yet he never abandoned a major research subject after encountering a roadblock. Instead, he set his research aside until he or others made the next breakthrough, after which he returned to attack formerly recalcitrant problems with new techniques.

This idiosyncratic research style required, of course, that he had an easy access to the newest mathematical literature. Euler thus found himself in the right place at the time, since groundbreaking results typically appeared in the proceedings of the great national academies, and these institutions exchanged their large folio volumes on a regular basis. Euler also gained information about the latest innovations through contacts in his wide network of correspondents, with some of whom he engaged in lively debates. In 1741, he accepted the invitation of Frederick II of Prussia to work at the Berlin academy, where he remained until 1766. There, he filled the mathematics section of the *Miscellanea Berolinensia* with his memoirs, just as he had done earlier when writing for the *Commentarii* of the Saint Petersburg Academy. After resigning from the Berlin Academy, he returned to Russia where he spent the last period of his life; although nearly completely blind, he continued to pursue mathematical research up until his death.

Euler composed a trilogy of works in algebraic analysis, or calculus in the sense of an extended symbolic algebra based on formal operations. Written in Latin, these texts systematically organized hundreds of important earlier discoveries beginning from the time of Leibniz and Newton. For historians of mathematics, they represent an essential turning point in the development of the discipline[2]. In fact, Euler regarded his *Introduction to Analysis of the Infinite* to be a prerequisite for reading his *Institutiones calculi differentialis* (Methods of the Differential Calculus, 1755) and his *Institutiones calculi integralis* (Methods of the Integral Calculus, 3 vols, 1768–70). He thus elevated the notion of functions, which was central for the first volume of his *Introduction to Analysis*.

FIGURE 3.4. Portrait of Leonhard Euler (1707–83) by Johann Georg Brucker (1756). As typical for the 18th century, Euler sits at his desk holding a quill, on his desk loose leaflets and a notebook. The size of the leaves suggests it is used for the redaction of a text. Photo by Bettmann / Getty images.

In its most general sense, a function expresses a relation between two quantities, for example, as often done in the sixteenth and seventeenth centuries, by means of tables of numbers. In the 1690s, Leibniz and Johann Bernoullli had employed this notion in a more specific geometric sense, namely, to describe relationships between different segments of a curve—such as abscissa, ordinate, arc length, subtangent—as these segments varied along the curve. While for them, as for earlier writers, the curve and its properties were the fundamental objects of their investigations, Euler introduced from the outset the notion of function of a variable quantity. He defined this as an "analytic expression composed in any way whatsoever of the variable quantity and numbers or constant quantities" (Euler, *Introduction*, vol. 1: 3).

Although curves and functions are closely related, they are by no means identical. Euler distinguished between continuous and discontinuous curves. The latter were composed of several continuous parts each of which was described by a distinct function, i.e., a distinct analytic expression (or formula). Euler used the notion of function to order the material of the two volumes. He gave a classification of functions divided into algebraic and transcendental (i.e., those which are not algebraic) such as exponentials, logarithms and trigonometric functions. In line with Leibniz and Bernoulli, he defined exponential functions as powers in which exponents are variable and then defined logarithms in terms of these. Namely, if $y = a^x$, x is the logarithm of y with base a. He dealt with "the transcendental quantities which arise from the circle," like sine, cosine, . . ., as functions, rather than as lines in a circle. The chapters on the transcendental functions completely changed the approach to these mathematical objects.

In volume 2 of his *Introduction*, Euler applied the "pure analysis" set forth in volume 1 to geometry, i.e., to curves whose main properties were derived from their equations. Like Gabriel Cramer in his *Introduction à l'analyse des lignes courbes algébriques* (Introduction to the Analysis of Algebraic Curves, 1750), Euler did not use infinitesimal methods to investigate these properties, although it would have been simpler to do so. His volume was devoted to the study of plane algebraic curves, specifically conics (described by equations of the second degree, i.e., circle, ellipse, parabola, hyperbola), cubics (third degree) and quartics (fourth degree). His classification of cubic curves completed and significantly simplified Newton's earlier classification.

Euler's attitude toward the infinite reflected his congenial approach to the calculus, in which he freely used both infinitely small and infinitely large quantities. Thus, in his calculations he operated with differentials understood as infinitely small quantities. Euler's most important tool for the representation of functions was to expand them as power series, which he treated simply like infinite polynomials by virtually ignoring all questions of convergence. Euler was convinced that all functions had a power series expansion, a conviction evidently based on experience. He was on safe ground, in fact, because functions given by analytic expressions do have such a representation. Euler's treatises stepped away from geometry entirely—diagrams, which were fundamental for Newton's calculations as well as Bernoulli's, are nowhere to be found in his *Methods of the Differential Calculus*. Euler's work thus symbolized a fundamental turn towards an algebraic approach to analysis; it established a conceptual framework for the increasingly autonomous field of analysis.

TOWARD AUTONOMOUS THEORETICAL DEVELOPMENTS

Thus far, we have taken a glimpse at the expansion of the calculus during the first decades of the century as well as some later, more systematic works that synthesized but also transformed earlier work. We have met some major figures, like Euler and d'Alembert, who worked in the fast-changing institutional context of (national) academies. These institutions generally offered an excellent research environment, replete with library facilities, publishing opportunities, and for a small but increasing number of *savants* the chance to compete for prizes. The problems posed—for instance, as prize questions by the academies, notably in Paris, Berlin, and St. Petersburg—aimed at solving open questions concerning a wide range of phenomena, such as planetary orbits, the shape of the earth, the causes of the winds, elasticity, fluid flow, etc.

In exploring this vast terrain, mathematicians sometimes used previously created analytical techniques, but they also found they needed at times to invent new mathematical methods. In some cases, they began to treat these methods, developed when solving higher-level problems, as mathematical objects in their own right. This led to a gradual shift in perspective, as researchers began to probe the theoretical properties of mathematical theories, which they studied for their own sake. In the following, we offer a brief survey of some of these more theoretical developments.

Applications of Analysis to Geometry

When studying natural phenomena, the mathematicians of this period typically thought in terms of the properties of bodies in physical space. To apply the techniques of analysis to such objects required having a strong mathematical grip on the geometrical properties of curves and (curved) surfaces. Mathematicians often conceived of curves as the paths of moving particles or objects, whereas they took the latter to be 3-dimensional bodies bounded by surfaces. These notions were, of course, long familiar from classical Greek geometry, but in that setting only a few idealized cases arose. In the second half of the eighteenth century plane curves were considered well known and lost part of their interest. Thus Newton and Euler had begun an investigation of higher-order plane curves, but the study of space curves began in 1731 with Alexis Clairaut, who investigated the properties of curves whose points lie on some surface other than a plane. The skew curve would be a nice example of such a curve. Clairaut called these curves of double curvature "as they were curved by themselves and by the surface on which they can be supposed applied" (D'Alembert, *Encyclopédie*, Art. "Courbe polygone").

Clairaut gave equations for some standard surfaces, including spheres, cylinders, and ellipsoids (which he used to study the shape of the earth). He could then investigate certain space curves by taking intersections of two such surfaces. This leads to a representation of space curves by two equations. With the help of the calculus, he then found tangents and normals to these space curves. Euler was also interested in the properties of surfaces. In a short treatise added to volume 2 of his 1748 *Introduction*, he investigated surfaces of solids, that is, 2-dimensional objects represented by equations with three variables, say $z = f(x, y)$. The analogues to tangent lines in this setting were tangent planes, but to find these one needed to measure rates of change in two directions. So finding the tangents to a surface described by $z = f(x, y)$ required calculating its partial differentials, and conversely, the solutions of partial differential equations represents families of surfaces.

An important class of these are the developable surfaces, such as the cylinder or the cone. Developable surfaces had considerable importance for mapmaking because one can flatten them out onto a plane without distortion (see also Chapter 2). In 1771, Euler published necessary and sufficient analytic conditions for a surface to be developable. Toward the end of the century, in 1795, Gaspard Monge also took up this subject in his *Feuilles d'analyse appliquée à la géométrie* (Worksheets of Analysis Applied to Geometry, first published in 1811). In this work, which he based on lecture notes, Monge was able to translate various properties of curves and surfaces into the language of partial differential equations. From today's perspective, one might describe Monge's *Worksheets* as descriptive geometry (see below) mixed with differential geometry (a term, however, only first introduced by the Italian geometer Luigi Bianchi in 1894). During the eighteenth century, one simply spoke of applying analysis or infinitesimal techniques to geometry, i.e., to the study of curves and surfaces. In the spirit of the Enlightenment, Monge explained many of his ideas by way of concrete examples. Thus, the problem of excavation and fill, which arose in building fortifications and involved transporting earth from one spot to another in the most efficient way, played an important role in motivating early work on curvature of surfaces.

Gaspard Monge's fame in the history of mathematics stems mainly from his reputation as the founder of descriptive geometry, a method for representing 3-dimensional objects in 2 dimensions (see Chapter 7 in Volume 5).

Monge was a singular figure in a century strongly dominated by analysis, insofar as he promoted geometry and analysis as well, regarding these two mathematical fields as interdependent. This viewpoint attracted noteworthy followers, but it gradually died out over the course of the nineteenth century. Monge never conceived of descriptive geometry as a new discipline. For him, it was a technical art; he developed its principles by systematizing earlier methods used by stonecutters and architects. Only after he began teaching these methods as part of the revolutionary curriculum, the discipline was established. Indeed as a staunch supporter of the French Revolution, Monge helped to found the

École polytechnique, where he taught up until the fall of Napoleon in 1815. The term "descriptive geometry" first appeared in a document written by Monge in September 1793, which described a teaching program for artists and workers in the newly founded secondary schools in France. The remarkable convergence between descriptive geometry and the revolutionary ideals of education led Joël Sakarovitch (1998: 167) to characterize Mongian descriptive geometry as a revolutionary discipline.

Systematic Study of Differential Equations

Mathematicians hit upon differential equations quite early, for instance, when exploring problems like isochrone curves or the shape of a sail under wind pressure (see above). They developed various integration techniques and other strategies in order to attack the equations that arose, such as separation of variables or substituting for the independent variable. In 1727–8, the young Euler became interested in the second-order differential equations that arise in connection with harmonic motion in resisting media. In these studies, he managed to reduce a whole class of second-order differential equations to first-order equations by a change of variables.

The idea of reducing the order of ordinary differential equations would prove important for mathematical posterity. Already a few years earlier, the Venetian Jacopo Riccati had applied this technique to handle the nonlinear differential equation that today bears his name.[3] However, Euler's memoir, published in the proceedings of the Academy of St. Petersburg in 1732, initiated the systematic study of ordinary differential equations (in one variable). Some ten years later, Euler tackled the problem of solving general linear equations with constant coefficients; he solved these in the homogeneous case, which he wrote: $Ay + By' + cy'' + \ldots + Ly(n) = 0$, where the coefficients are constants. By 1750, he could also solve the nonhomogeneous case (where the right hand side of the equation is not 0) by reducing the order of the equation systematically. In 1762, Lagrange took up the more general case, where the coefficients of the linear differential equation are no longer constants, but he was unable to find a general method of solution. Others afterward searched for such a method, but they only obtained partial results.

Since most physical problems involve more than one variable, mathematicians realized they needed to extend the differential calculus to several variables. This began rather early in the century in the context of studies on families of curves. Partial differential equations involved as unknown a function of several variables and partial differentials with respect to those variables. Leibniz and later Johann Bernoulli were able to calculate partial differentials by applying the standard rules when one of the variables remains constant. They introduced no special symbols for partial differentials, however, and the term "partial differential equation" appeared only towards the end of the century with Nicolas de Condorcet. Euler's *Methods of Integral Calculus* (3 vols, 1768–70) contained solutions of the most elementary differential equations and all basic integration results. These texts thus offered a comprehensive catalogue of the available solutions of ordinary and partial differential equations. D'Alembert's earlier prize essay (Berlin 1746) on the cause of the winds is considered one of the first attempts to make consistent use of partial differential equations.

We will turn instead to discuss briefly the vibrating string problem, one of the most hotly debated topics in the second half of the century. Its lasting fame in the history of mathematics has much to do with later developments, as the issues at stake were only resolved in the nineteenth century through the work of Joseph Fourier (see Chapter 3 in Volume 5). The key point in question concerned the nature of an "arbitrary" function and its expansion in trigonometric series.

Vibrating String Problem

In the late seventeenth century, members of the Royal Society in London witnessed the performance of various musical experiments, which also caught Brook Taylor's attention. Not only was Taylor experimenting himself with paper instruments designed to embody a music theory, in 1713, he set forth the first mathematical description of a vibrating string (Wardhaugh 2009: 642). If one plucks a string, making it vibrate, the vibrations of the string form standing waves. Taylor adopted the hypothesis that the vibrations were very small, from which he obtained a formula for the fundamental frequency of a stretched vibrating string (in its oscillations the string produces a single arch). Today, we may recognize in this formula the wave equation[4] involving second order partial differentials, but there is no evidence that Taylor himself had any notion of those equations.

Toward the middle of the century, several investigators took up the problem again in a series of papers published in the Memoirs of the Berlin Academy between 1747 and 1753. The search for the general shape of a vibrating string, i.e., for all possible motions, began with d'Alembert. In a paper published in 1749, he proposed to show that an infinity of curves were possible waves of propagation that satisfied the conditions of the problem (beyond the sine curve, already identified by Taylor as a solution). As the position of a point on the string varies both with its abscissa (x) and with time (t), its movement will be a function y of these two variables. Using partial differentiation and Newton's laws, d'Alembert derived the equation for y (written here in modern notation):

Insert 2

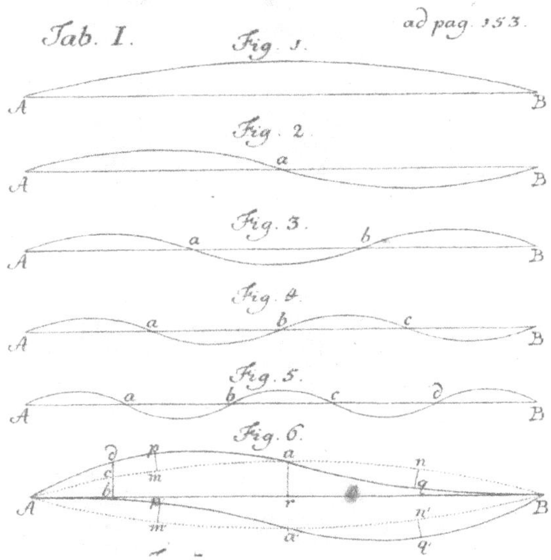

Notes:

Vibrating string as represented by Daniel Bernoulli in Mém. Berlin 1753 (1755), table I, figs 1–5, p. 353.

A stretched string AB can perform uniform vibrations in an infinity of ways. These are characterized by the number of loops that the string can form during its vibration.

Fig. 1 shows the fundamental tone (one loop, the vibrations are the slowest and produce the fundamental tone).

Figs 2–5 show overtones, fig. 2 the octave of the fundamental tone (two loops, one node in the middle of AB, the vibrations are doubled).

When the string forms 3, 4, or 5 loops, with 2, 3, or 4 nodes, at equal distances, then the vibrations are multiplied by 3, 4, or 5 (figs 3–5).

$\frac{\partial^2 y}{\partial t^2} = \frac{\partial^2 y}{\partial x^2}$, where ∂ is the notation for partial derivatives, later introduced by Lagrange.

D'Alembert was mainly concerned with a general mathematical solution to this partial differential equation and took less interest in the physical or musical phenomena it described. For him "the application of mechanics to geometry sometimes helps to discover purely geometrical truths" (*Histoire de l'Académie royale, Berlin*, 3, 1747 (1749): 230). He was able to solve the above equation, but the general equation of the solution curve included two still unknown functions. D'Alembert then discussed the special case where the string was in a position of equilibrium (i.e., stretched out in a straight line) at the beginning of the oscillation; furthermore, he required so-called boundary conditions, namely that the string remain fixed at the two endpoints $x = 0$ and $x = l$, where l is the length of the string. He then generalized this solution to cases where:

(1) the initial position of the string is not a straight line but any curve given by a function $f(x)$, and

(2) the motion does not start from rest, i.e., the initial velocity is not zero, but given by another function $g(x)$.

He only required of the functions f and g that they be given by a single analytic expression (to ensure they are twice differentiable). This, he maintained, ruled out the possibility that the initial curves corresponded to functions only defined piecewise, as this situation would "surpass the power of known analysis" (quoted in the English translation by Struik 1986: 360–1). D'Alembert's conditions on the allowable initial solutions were thus dictated by purely mathematical considerations.

In a paper published shortly afterward, Euler objected to d'Alembert's specification of the "arbitrary" functions (f and g) included in the general solution of the wave equation. Instead, he imposed no constraints on the initial form of the vibrating string. Euler's method of solution otherwise close to d'Alembert's, differed with it sharply when it came to the nature of functions that could be admitted as initial curves. In particular, Euler allowed for possibly "discontinuous functions," thus curves formed from parts of different curves or even curves drawn freehand in the interval $0 < x < l$, which d'Alembert excluded from geometry. For Euler, this was instead a great step forward, as "considering such functions as are subject to no law of continuity opens to us a wholly new range of analysis" (Euler to d'Alembert on December 20, 1763 quoted by Kline (1972: 507)). By now, Euler was well aware that the narrow definition of function in the tradition of algebraic analysis was insufficient to describe the phenomenon of the vibrating string and that a broader concept of function was needed.

Daniel Bernoulli, whose father Johann had earlier encouraged him to investigate the vibrating string, had in the meantime acquired a solid understanding of the physical phenomenon. After reading d'Alembert's and Euler's papers, he reacted strongly, insisting that all possible initial curves were representable as an infinite series of trigonometric functions. He further asserted that all subsequent motions would be given by an expression of the form:

$$y(t,x) = \sum_{n=1}^{\infty} a_n \sin\frac{n\pi x}{l} \cos\frac{n\pi x}{l} \quad (*).$$

In Daniel's words, d'Alembert's and Euler's calculations "certainly contain all that analysis can have at its deepest and most sublime, but [. . .] show at the same time that an abstract analysis which is accepted without any synthetic examination of the question under discussion is liable to surprise rather than enlighten us" (Struik 1986: 361). The notion of "synthetic examination" as used here served as a plea to focus the debate less on the mathematics and more on the underlying reality of physical vibrating strings. Bernoulli drew attention to the fact that many modes of a vibrating string could exist simultaneously and that the string then responded to the sum or superposition of all the modes. He thereby gave an explicit formulation of the so-called principle of superposition, a previously unknown property.

For Bernoulli, relying solely on physics, every motion corresponding to any initial curve was thus no more than a sum of sinusoidal periodic modes (as in equation (*) above). Euler, who had engaged in lively discussions with Daniel when both worked in St. Petersburg, replied immediately to his friend with objections. He denied (as did d'Alembert) that all possible motions of the string could be expressed by such sum, though he saw that Bernoulli's solution would be legitimate if every function could be represented by an infinite trigonometric series. However, neither Euler nor his contemporaries were convinced that this was true in general, and so the debate between d'Alembert, Euler and Bernoulli continued to rage through two more decades. In 1759, the young Turinese mathematician Lagrange entered the contest, and he alone referred to the experiments undertaken at the turn of the century by the French Academician and acoustician Joseph Sauveur, who had then studied standing waves. It remained to mathematicians of the next century to struggle with questions regarding those functions representable by an infinite trigonometric series (see Chapter 3 in Volume 5).

This debate between three giants of eighteenth-century mathematics was also significant in that it reflected three different perspectives. D'Alembert, as a mathematician, was mainly interested in the integration of partial differential equations. Daniel Bernoulli, a traditional physicist, was interested in the phenomenon itself and invited fellow mathematicians to join him in making observations of vibrating strings. Euler, a mathematical physicist, was the personification of the strong alliance between mathematics and mechanics, as often emphasized by historians of mathematics. He clearly realized that the arbitrary functions, which entered in the solution, posed completely new problems for the analysis of his time (Jahnke 2003: 127).

Calculus of Variations

In the aftermath of work on the brachistochrone problem, mathematicians began investigating similar problems involving curves that minimize a particular quantity. These included isoperimetric problems[5] as well as the problem of finding the shortest curves that connect two points on a surface, all of which belong to an early stage of what Euler in 1756 called the calculus of variations. Twelve years earlier, Euler (see Figure 3.5) effectively launched this new branch of calculus in *Methodus inveniendi lineas curvas maximi minimive proprietate gaudentes* (The Art of Finding Curved Lines, which Enjoy Some Maximum or Minimum Property). This work created general methods for handling over one hundred problems, placed in order of increasing complexity. A typical problem from this new field takes the following form: to determine a function $y = y(x)$, which appears in the integrand of an integral expression, $I = \int_{x_1}^{x_2} f\left(x, y, \frac{dy}{dx}\right) dx$, so as either minimize or maximize the value of I. In the brachistochrone problem, for instance, one seeks a curve that realizes the minimum time of descent for a heavy particle that moves along the curve between two given points. Thus, here the integral I represents time, which must then be minimized.

Euler replaced the integral by a finite sum (in true Leibnizian style), then varied one or more arbitrarily selected ordinates of the minimizing or maximizing curve, and finally calculated the variation of the integral. To attain a minimum or a maximum, the variation had to vanish (thus, equal zero, according to the rules of the calculus). He then obtained, following some obscure intermediate steps, a differential equation which had to be satisfied by the curve. Euler was thus able to obtain a necessary condition in the form of a second-order differential equation, which $y(x)$ must satisfy in order for I to assume an extreme value. This condition, today known as the Euler-Lagrange equation in the calculus of variations, remains a fundamental result of the subject.

The young Lagrange had immersed himself in these kinds of problems, and on August 12, 1755, he wrote Euler a letter that left a deep impression on the latter. Lagrange was less than sympathetic with the semi-geometrical methods of Euler, and so he approached variational problems by using purely analytical means. Instead of varying selected points on the curve, Lagrange considered variations of the entire curve. Moreover, in analogy with the symbol *d,* used to denote tiny differences in the calculus, Lagrange introduced a special notation δ for the variation of the curve. When applied to the integrand of I, the new curve $y(x) + \delta y(x)$ induces a transformation of I, which produces the variation δI. To obtain the minimizing or maximizing curve $y = y(x)$, Lagrange showed that this δI had to vanish, a condition he found almost effortlessly. This confirmed the same necessary condition for an extreme value that Euler had derived in 1744, thereby bestowing confidence in this new branch of analysis.

Lagrangian Algebraic Analysis

While certain parts of the calculus acquired substantial autonomy, the larger question concerning the solidity of its foundations remained open. An important

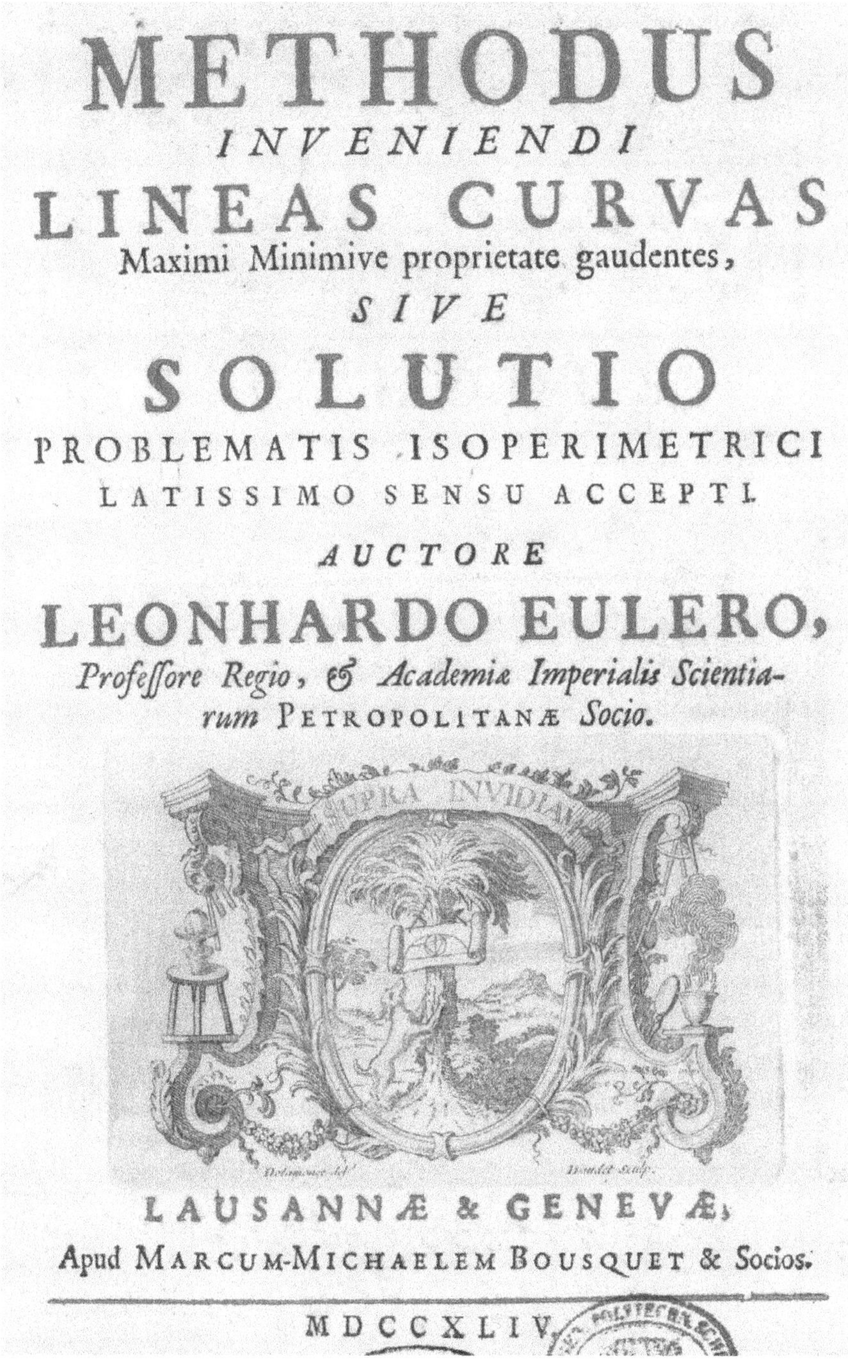

METHODUS
INVENIENDI
LINEAS CURVAS
Maximi Minimive proprietate gaudentes,
SIVE
SOLUTIO
PROBLEMATIS ISOPERIMETRICI
LATISSIMO SENSU ACCEPTI.
AUCTORE
LEONHARDO EULERO,
Professore Regio, & Academiæ Imperialis Scientiarum PETROPOLITANÆ *Socio.*

SUPRA INVIDIAM

LAUSANNÆ & GENEVÆ,
Apud MARCUM-MICHAELEM BOUSQUET & Socios.
MDCCXLIV.

FIGURE 3.5. Title page of Euler's treatise *Methodus Inveniendi Lineas Curvas maximi minimive proprietate gaudentes* (1754). On the tree the picture of a cycloid is hung. Photo courtesy of ETH-Bibliothek Zürich, Rar 4999. Public Domain.

landmark on the way to securing them was Lagrange's *Théorie des fonctions analytiques* (Theory of Analytic Functions, 1797), a work based on his lectures at the recently founded École polytechnique in Paris, where Lagrange had settled. His idea was to base the calculus on algebraic analysis, that is, that part of mathematics resulting from the enlargement of algebra to include calculus-related methods and functions (Fraser 2008). In Lagrange's view, to place the calculus on rigorous foundations required eliminating all reference to infinitesimals, a goal he failed to accomplish completely. Starting with Euler's notion of a function as an analytic expression, $y = f(x)$, he introduced the "derived function" (or derivative, denoted $f'(x)$ by Lagrange). He defined this formally as the coefficient p of the linear term in the power series expansion $f(x + i)= f + pi + qi^2 + ri^3) \ldots$, where $p, q, r, \ldots$ are functions of x independent of i. Lagrange then showed that this power series, whose existence he never questioned, was identical to the Taylor series:

$$f(x+i)=f(x)=if'(x)+\frac{i^2}{2}f''(x)+\frac{i^3}{2.3}f'''(x)+\ldots$$

Thus, the derived function $f'(x)$ corresponded to the quotient $\frac{dy}{dx}$. The central step here was the passage from one function to another, called the "fonction derivée," as it was derived from the first. As a result, the emphasis on the study of curves gradually shifted to the study of functions, thereby paving the way for the algebraization of analysis. What Augustin Louis Cauchy later termed the "generalities" (see Chapter 3 in Volume 5) of such an algebraic approach provided fuel for research that would remove the ambiguities that emerged.

The more theoretical developments sketched above were not considered autonomous branches of mathematics but rather as parts of higher mathematics ("mathématiques sublimes") and taught as such. The theories of ordinary and partial differential equations, for example, did not constitute a special field but rather a part of integral calculus broadly defined. Even if historians have aptly characterized this period as the century of analysis, a number of other important developments took place, especially in algebra.

In fact, during the eighteenth century algebra was hardly distinguished from analysis, as witnessed by Lagrange who used both terms synonymously. Work in algebra often had close ties with analysis, as with the so-called fundamental theorem of algebra, which states that every polynomial equation of degree n with real coefficients has n solutions, either real or complex. This is an example of a pure existence theorem, since it only claims the equation has a certain number of solutions but says nothing about how one might find them. Moreover, some may be rather mysterious, being no longer real but imaginary (or complex) because they depend on the number $\sqrt{-1}$.

Complex numbers can be written $a+b\sqrt{-1}$, where *a*, *b* are real numbers. They have a curious history, having first barged into mathematics in the Italian Renaissance when algebraists noted their role in solutions of cubic equations (see Chapter 3 in Volume 3). Even in the first decades of the eighteenth century, these imaginaries remained quite unfamiliar to many mathematicians. Still, they turned out to be highly useful, for example, when integrating rational expressions via the method of partial fractions. One can use this method, for instance, to calculate $\int \frac{dx}{a^2 - x^2}$, simply by rewriting the integrand as $\frac{A}{a-x} + \frac{B}{a+x}$, where *a* and –*a* are the two roots of $a^2 - x^2 = 0$. This then integrates immediately (as the sum of two logarithms). More generally, if the denominator of the integrand is an arbitrary quadratic, $ax^2 + bx + c$, its linear factors might well be complex. This then raised a natural question: can one always decompose a general polynomial into linear and quadratic factors by means of its roots? That question, though, amounts to just another way of formulating the fundamental theorem of algebra.

D'Alembert came up with a proof in 1746, but this was incomplete. In 1749, Euler produced a correct proof for cubic and quartic polynomial equations, but his argument failed for polynomials of higher degree.[6] Then, in 1770, Lagrange published a paper that attempted to understand why the method that succeeded for degrees three and four broke down for higher-degree equations. As an unanticipated result of this inquiry, Lagrange introduced a completely new approach to the theory of algebraic equations involving permutations of their roots. As it happened, this provided the foundation for major new developments in algebra during the early nineteenth century (see Chapter 3 in Volume 5).

CONCLUSION

Apart from a host of specific results in nearly all branches of mathematics—theory of equations (or algebra), series, number theory, properties of curves, surfaces and solids, etc.—the main mathematical invention of the eighteenth century was a new field of mathematics, namely analysis. Leibniz was well aware of this when he began mapping the outlines of the field early in the century. Within it, he saw himself as occupying the central position, while he assigned positions in the field to his epigones, thereby enabling them to shine in their various public jousts. D'Alembert in the 1750s described the new analysis as a specific type of problem solving method that used equations. As such, it was synonymous with algebra, but it was also a tool for invention and an art of abbreviating.

Among the modes of mathematical invention current in the eighteenth century, we have highlighted the disputes that dominated the early circulation of the calculus, fueled by competition among individuals or countries.

Controversies were part of a mathematical culture rooted in a problem solving tradition, where the few individuals able to answer new questions were honored as heroes. The newly created learned journals were to strengthen this trend and gave the disputes among mathematicians greater publicity. More generally, the new form of journal articles proved to be fruitful for the circulation of the new analytical methods, especially on the European continent. On the other hand, the ordering of single results in voluminous treatises has proved in Euler's hands to lead to a complete reorganization of the nascent field of infinitesimal analysis—an innovation in itself but also an incentive to uncover hitherto unknown questions. The solution of complex physical problems, such as the cause of the winds, often posed as prize questions by the European Academies needed more sophisticated techniques—another source of invention. Mathematicians had a strong faith in their methods and pushed forward although the foundations of infinitesimal analysis came sporadically under attack. Toward the end of the century, a strong bend towards algebraization, away from the Eulerian geometrical language was considered (especially by Lagrange) as a way to solve foundational questions, however without success. These came to occupy nineteenth-century mathematicians, and among them Augustin Louis Cauchy was prominent (see chapter 3 in volume 5).

CHAPTER FOUR

Mathematics and Worldviews

Newtonianism and Probabilistic Worldviews

DAVID BELLHOUSE AND JIP VAN BESOUW

The Enlightenment is classically seen as a period in which a so-called "rational" order of knowledge came to prevail. This rational order of knowledge informed and bolstered the religious worldviews of some but also inspired others to form secular worldviews. The changes in the order of knowledge started in Europe and had its roots in a number of societal and economic changes. Mathematical science was an important element of these changing worldviews. The structure of mathematical reasoning was often invoked as a model for understanding the real world quantitatively or qualitatively, and the progress made in understanding the universe definitely influenced old and new worldviews alike. Over the course of the century, however, a slow but definite delineation or even separation of the realms of metaphysics and mathematics set in.

At the beginning of the century, for example, mathematics still played a central role in upholding Gottfried Wilhelm Leibniz's argument of a pre-established harmony of the universe. At the end of the century, however, Immanuel Kant would claim that mathematical and metaphysical knowledge are separate forms of knowledge and that mathematics could not be used to prove the existence of God or the harmony of the universe (see Chapter 7). A similar evolution occurred in that other important eighteenth century group of worldviews, Newtonianism. In the first part of this chapter,[1] we discuss a variety of ways in which Isaac Newton's legacy was interpreted and made serviceable for harnessing worldviews. The early English reception of Newton helped to validate providentialism, but by the end of the century, Newton's legacy would generally be associated with Pierre-Simon Laplace's areligious view on the world, one determined by mathematical law alone.

In parallel with this evolution, another important concept, that of probability, acquired new dimensions, too. At first associated only with (theoretical) games, probability became an independent concept during the eighteenth century. The concept was negatively defined. Whereas God as the creator of the universe would have perfect knowledge about the nature or about the causes of things, the essential incompleteness of human knowledge translated as probability. Approaching the nineteenth century, Laplace's secular universe still featured this definition, "probability is relative, in part to this ignorance, in part to our knowledge," but any divine intervention or miracle to supplement an unknown cause was now regarded "an illusion of the mind" (Laplace 1814). Instead he appealed to a continual, though infinite progress of the human mind towards "an intelligence sufficiently vast to submit [all] data to analysis," and capable of tracking all movements of a mechanistic universe. This probabilistic worldview is the topic of the second part of this Chapter.[2] It shows how domains of nature and society that could not yet be fully grasped through laws and determinism, could still, and increasingly so, be made amenable to the calculus of probabilities. This process continued well into the nineteenth and twentieth centuries (see Chapter 5 in Volume 5 and Chapter 1 in Volume 6).

THE MANY FACES OF NEWTONIANISM

Isaac Newton's *Philosophiae naturalis principia mathematica* of 1687 rocked the intellectual world and attained an iconic status in the eighteenth century. Newton's popularity went upwards in parallel with his own career: in the decades after the first publication of the *Principia* Newton took on various political positions and was elected President of the Royal Society of London in 1703. His work, at the same time, increasingly reached wider audiences, first through the publication of the *Opticks* in 1704, but mostly because of the second edition of the *Principia* in 1713.

Throughout the century, Newton's mathematico-experimental approach to the natural world was saluted as an exemplary method for scientific investigation by many, but was also the primary target for those who wished to restrict the domain of mathematics. Through Newton, "mathematics" turned into one of the buzzwords of eighteenth-century intellectual culture. As an icon, however, Newton stood for much more than just a particular scientific methodology. A mention of Newton or the "Newtonian philosophy" could, depending on the context, also indicate certain cosmological, theological, and even political positions. Furthermore, a diverse range of mathematicians, philosophers, and entrepreneurs in the eighteenth century adopted "Newton" as brand name without necessarily having much in common with the man himself.

Today, the *Principia* is known as one of the most distinguished texts in the history of the sciences, on par with only a few others such as Copernicus' *De revolutionibus orbium coelestium* and Charles Darwin's *On the Origin of Species*. The *Principia* set out Newton's laws of motion and universal gravitation and show how these explain the motions of the heavenly bodies. However, the modern mathematical expression of these laws are nowhere to be found in Newton. Instead, they are the result of a historical process of reformulation, discussion and synthesis leading up to analytical mechanics (see Chapter 5 for more details). Indeed, anyone who has ever opened a copy of the *Principia* will have seen that it is a perplexing book that does little to accommodate its readers. Due to certain distinctive choices and innovations in structure, techniques, and notation, even Newton's most talented peers had difficulties to plough through the *Principia*. Over the course of the eighteenth century, by which time Newton's techniques and style had become increasingly outdated, much of the book would be increasingly unintelligible to the new generation of mathematicians (Guicciardini 1999: 107–8).

The fact that Newton attained stellar status in the eighteenth century is thus in need of an explanation, for how could such a book become a success? A major condition for that success was, of course, that the people who initially managed to study the *Principia* thought highly about it. This was certainly the case. British mathematicians and astronomers were quick to pay tribute to Newton, leading Continental mathematicians such as Christiaan Huygens and Leibniz were very favorably impressed, and the book was read avidly in mathematical centers such as Basel and Paris. Most of its initial readers were impressed by the many interesting mathematical problems and the far-reaching mechanical and astronomical models. These were found especially in Book 3 of the *Principia*, called the System of the World, which explained the terrestrial and celestial motions using the law of gravity (see also Chapter 5).

But the *Principia* did not formulate a worldview. Certainly in the first edition, there are few programmatic statements on how to do or interpret science and little trace of a comprehensive worldview. Newton had even explicitly refrained from addressing the causal processes underlying the posited force of universal gravitation (Cohen 1971: 145–61). Compared to contemporaneous books on natural philosophy, the *Principia* was also particularly silent on topics such as its method of investigation and God's relation to the physical universe. Although they immediately realized Newton had unsettled the leading cosmological model of the Cartesians, the likes of Huygens and Leibniz did not feel that Newton had much of an alternative to offer.

Nevertheless, readers found what they took to be suggestions about worldviews and were quick to march ahead where Newton had stopped. They addressed the topic of the cause of universal gravitation and many other issues found in the *Principia*. Newton's work generally was taken up by many different authors and developed in many different directions throughout the eighteenth century. As a result, Newton's legacy was not a coherent philosophical program or a well-defined worldview. Rather than that, terms such as "Newtonianism" should be understood as a name for a fuzzy group of beliefs and practices changing with the contexts in which people read Newton or, most of the time, talked about Newton without reading his works. It is exactly this feature, however, that assures that Newtonianism offers a fruitful window onto the connections between mathematics and worldviews. Here, we focus particularly on debates on causation, theology, and cosmology, the detachment of post-Newton natural philosophy from those debates, and images of Newton created in the Enlightenment. As a consequence, little will be said about Newton's *Opticks* (1704), which was an important driving force for the eighteenth-century sciences in its own right, but not as directly connected to mathematics and worldviews as the *Principia*.

Newtonian Providentialism

Natural philosophy was an enterprise that readily spoke to theology. The philosophers of Newton's own generation were generally committed to a Christian worldview: not to do so was still a reason to be shunned. This meant, however, that they had to align conventional ideas about God's activity in the world with an informative account of how causal processes took place *naturally*. The bulk of Newton's peers conceived of these natural causal processes in terms of something often called "corpuscularism." On this view, the fundamental entities of the physical world were small material particles or bodies. Physical processes, sometimes including mental processes, were subsequently thought of in terms of powers or forces residing in matter, and the motions of and mechanical impacts between pieces of matter. At least since René Descartes, one of the shared goals of natural philosophers was to describe the natural phenomena, or at least those with some regularity, in terms of laws of nature.

Opinions differed strongly, however, on how and if the described forces or laws could cause the natural phenomena. To see the predicament of Christian natural philosophy, let us grant that the universe is ordered by laws of nature such as those of the *Principia*. If we now accept that these laws represent the causes of the phenomena, one could expect such laws hold universally, certainly when they can be formulated as mathematical regularities. But how can this be reconciled with the idea that God is the first cause and guardian of the world and as such superior to the natural processes studied by natural philosophy? How can one reconcile a mathematically ordered world with the idea that God can act freely in the world, or with the Biblical stories about miracles? Does this mean that God has to suspend the order of nature whenever he wishes to perform miracles? Or, rather, are the laws of nature in fact not causal entities but just mathematized descriptions of how God, the real cause of the phenomena, chooses to act most but not all of the time? If so, does that mean we actually cannot expect heavy objects always to fall down?

The role of such questions in natural philosophy around 1700 should not be underestimated. Leibniz, for example, addressed three of the principal philosophical issues flowing from the predicament above in his *Theodicy* of 1710: the being of God, the paradox of free will, and the question of why a fundamentally good and just God, let evil exist in the world. Newton, however, remained an outsider to the explicit debates on the topic (Guicciardini 2018a: 133–4). This was not because he did not believe such debates to be important: Newton privately thought and wrote about the meta-processes taking place in nature and gave God a major role in such thinking, but seems to have believed that in the end we cannot second-guess God's means of acting. And although Newton wrote extensively on religious and theological matters—Newton held a set of deep but idiosyncratic and heterodox religious positions, among other things forcefully rejecting the doctrine of the Trinity—most of his thoughts on the matter remained unpublished and his religious thought had little impact on the later eighteenth century (Iliffe 2017).

Thus, the first edition of the *Principia* remained silent on most religious and philosophical issues. Nevertheless, the book contained a theory that directly fed into existing debates. Newton introduced a number of different "forces" in mechanics and astronomy and gave his readers plenty of reason to believe that he thought of these gravitational forces, elastic forces, and impressed forces not just as mathematical shorthands but as real causal factors. But, as he clarified with regards to the force of gravity in the second edition of the *Principia*, he did not "feign hypotheses" about the ontology of gravity as he had no way of finding empirical support for such an ontology. Even though, somewhat contradictory, his famous Query 31 to the *Opticks*, introduced in 1706, contained a discussion of possible causes of gravity, Newton never clearly supported one particular explanation (Guicciardini 2018a: 209–10).

Exactly the fact that Newton did not take a clear position allowed his readers to take whatever they could find in the *Principia* and use it to break new ground. Arguments from the *Principia* entered debates on causality and theology in 1692, when the classicist and clergyman Richard Bentley made several references to Newton's work in his *The Folly and Unreasonableness of Atheism*

Demonstrated. This book collected the sermons Bentley had given as Boyle Lectures in 1692. In these sermons, Bentley set out to prove that atheism, materialism, and deism were wrong and that, instead, there was a single omnipotent and active God, as Christianity asserted. Newton's law of universal gravitation played an important role in these lectures and the arguments put forward by Bentley were rehearsed and refined over the next years by other Anglican clergymen such as William Whiston and Samuel Clarke, whilst they were countered or turned on their head by freethinkers such as John Toland and Anthony Collins (Ducheyne and van Besouw 2021).

What Bentley argued against was the view that all natural phenomena could be explained in terms of matter and motion—what we could call materialism—and that God's providence was not needed. Bentley argued that various complex phenomena, among which human thought and the complexity of animal and human bodies, showed that there must in fact be an intelligent creator. This, however, did not show that this creator was still active in the world. To argue for that point, Bentley turned to gravity. The leading conception of planetary motion before the *Principia* was that it emerged due to the mechanical collisions of large numbers of particles in motion, not due to some mysterious specific force. Particularly, Descartes had argued in his *Principia philosophiae* (1644) that the heavens were filled with subtle matter whirling around in vortices and that the combined impact of these little particles explained the planets' orbits around the Sun. Now Newton's *Principia* offered decisive quantitative astronomical arguments against Descartes' theory.

Bentley, who was well informed but not a mathematician or a philosopher himself, took a different approach, developing a particular interpretation in which God was directly and continuously infusing matter with gravitational powers. Specifically, Bentley pointed to Newton's argument that a world filled with matter everywhere would not be able to account for the specific gravity of different bodies. Although this was arguably one of Newton's weaker arguments against the vortex theory—the argument was significantly toned down from the second edition of the *Principia* onwards (Newton [1726] 1999: 810)—it was this argument that Bentley and others took to be decisive. For it proved, they believed, the existence of a vacuum, that is, the existence of empty space. Given the existence of empty space, they argued, gravity could not be caused mechanically or materially. Thus, the only explanation of the planetary motions left, Bentley believed, was that there was an immaterial cause, God, which continuously acted on each particle of matter in the universe. That is, Bentley and those who followed him believed, gravity demonstrated God's direct providence.

The use of such arguments in the 1690s cannot be understood without taking account of a much broader use of providentialist arguments. This rise was part of the political and cultural upset caused by the invasion of England by the stadholder of the Netherlands, the future King William III of England, in 1688. His supporters justified the takeover on religious grounds: the Protestant William's fleet had profited from a sudden change of winds to sail past the navy of the Catholic King James II. Surely, the underlings of William argued, this was a sign of God intervening on behalf of the right religion? It is against this background of legitimizing the rule of William III and the political and religious upheaval caused by it that Bentley, an upcoming man in the Church of England, was casting the *Principia* as supporting providentialism.

Such was certainly not the only possible reading of the *Principia*, which had actually been dedicated to James II, in 1687 still the undisputed king. Moreover, with the very few religious references in the text itself, one wishing to test the orthodoxy of Newton's work would have been drawn to the poem by Edmund Halley's that opened the book by praising Newton's achievements. Halley, who had the reputation of being irreligious, tacitly embraced classical and particularly Epicurean themes in his poem such as an eternal world consisting of atoms, swerving around by some mysterious powers. Such associations between the *Principia* and Epicureanism, which quickly lost all its fashionability after the events of 1688, were noted by Leibniz and, much later, Immanuel Kant, and were certainly not that far-fetched (Schliesser 2013). Halley's case illustrates the dire consequences of supporting a philosophical view that ended up on the wrong end of politics. In 1691, Halley, an outstanding astronomer and mathematician, was rejected for a chair of astronomy at Oxford. At that occasion Newton did nothing to help Halley out, even though Halley had in many ways been the driving force behind the publication of the *Principia*. Even worse for Halley, Newton in fact successfully endorsed one of his opponents (Albury 1978).

Disputes and the Closing of the Ranks around Newton

The case of the Oxford chair of astronomy nicely shows that political and religious views could play an important role, but by no means always did so. The chair in fact passed to the Scot David Gregory, who previously had held a chair in mathematics in Edinburgh, since 1683. In Edinburgh, Gregory had the reputation of a somewhat irreligious man, particularly through his association with the professor of medicine, Archibald Pitcairne. Besides being unorthodox with respect to religion, these men were also political outcasts: both were defenders of the disposed James and seem to have supported the Tories. This put them diametrically opposed to Bentley, Clarke, and Whiston, who were members of the clergy, and Whigs, supporters of the new political order, and, through all these positions, close to Newton's own public political stance (Guerrini 1986). Still, these putative differences did not prevent Newton from becoming closer to Pitcairne and especially Gregory.

Likely due to their political position, both Scots took their work elsewhere, Gregory to Oxford and Pitcairne to Leiden. In their near simultaneous inaugural lectures at their new universities, held in April 1692, both argued strongly in favor of applying "mathematical methods" in their new fields of teaching, astronomy, and medicine, respectively. While rather straightforward in the former, we should take this mathematization of medicine with a pinch of salt. Pitcairne incorporated some insights from geometrical optics and hydrostatics into his medical teaching, but mathematically speaking this did not amount to more than crude attempts to quantify fluid flows in the human body. Nevertheless, in the 1690s and 1700s, many students of Pitcairne and Gregory joined them in their vocal support for Newton's *Principia*. In natural philosophy, these figures argued mostly in favor of mathematical or experimental method and against the Cartesian vortex theory, which they denounced as a speculation. Most importantly, in the 1700s we can see the rise of something akin to the idea that Newton had a particular approach to natural philosophy, a way that could be applied beyond what could be found in Newton's own writings. This is illustrated by George Cheyne, a would-be physician and student of Pitcairne's. In the first decade of the century, Cheyne started to dabble in fluxional calculus, wrote on the connection between natural philosophy and religion staking out a variant on Bentley's position, and argued that medicine should be modelled on the *Principia*.

Among the students of Gregory in particular, we find a number of the most important British mathematicians of the time, among whom was John Keill. Keill is best known for his role in the priority dispute over the calculus in the 1710s, wherein he took a leading role in accusing Leibniz of plagiarizing crucial insights. Newton's work on the topic dated mostly from the 1660s, but at the turn of the century most of it remained unpublished (see Chapter 3). Although Newton and Leibniz had happily corresponded in the past, even after Newton's initial accusations of plagiarism referring to Leibniz's work of the 1680s, by 1710 their personal relations had deteriorated to a level of nasty bickering more readily associated with today's tabloids than with exchanges between great minds. While Leibniz asserted his claim for priority personally, Newton only wrote anonymously on the subject. He left most of the dirty work to others, whom he allowed to publish parts of his extensive manuscript material on the fluxional calculus (Guicciardini 1999: chapter 3 and chapter 7).

The second edition of the *Principia*, of 1713, staked out a clearer position than the first on a number of issues that relate to worldviews. The most prominent additions to the book were a new preface by Roger Cotes, one of Newton's most important interlocutors, and the "Scholium Generale" that Newton added at the end of the book (Newton [1726] 1999: 385–99, 939–44; cf. Smeenk and Schliesser 2013). Both texts virulently attacked the Cartesian vortex theory, asserted that the cosmology of the *Principia* led to the proper conception of an almighty God who actively presided over his creation, and argued that the *Principia* was guided by the right scientific methodology. Most importantly from our perspective, is that Cotes deliberately presented "the Newtonian philosophy," a phrase with virtually no currency before 1713, as a proper alternative to the well-established Cartesian and Aristotelian philosophies. In this respect, Cotes' preface could be said to be the first proclamation of the Newtonian philosophy as a worldview. Both mathematics and experiment played an essential part in positioning this worldview. Particularly, Cotes argued that Newton had mathematically *demonstrated* the existence of the force of gravitation while others such as Descartes had built their cosmologies on metaphysical speculations. As Newton himself added in the General Scholium, he did "not feign hypotheses" because such hypotheses had no place in the proper, new way of philosophizing, which was to proceed via a carefully regulated process of induction (van Besouw and Ducheyne 2021).

In refuting the Cartesians on their method, Cotes and Newton also sniped at the philosophical ideas of both Spinoza and Leibniz. Although the similarities between the three are easily overestimated, Leibniz and Spinoza did agree with Descartes that the geometrical method held the key to philosophical knowledge. What they understood by this method, however, was the proving of philosophical propositions by deduction from self-evident first principles. Rather than just referring to deductive or constructive procedures, Cotes' Newtonian mathematical method was also about quantification and generalization. It was used first to turn experimental findings into laws of nature, and second to explain particular phenomena through comparisons with the predictions of these mathematically formulated laws, a process explained by Newton in the somewhat cryptic "rules of philosophizing" he inserted into the second edition (Ducheyne 2012: 107–20). According to Cotes and Newton, true principles were only to be found via this experimental-mathematical method. Obviously, combining experimental and mathematical methods had been commonplace among natural philosophers of the seventeenth century, too. Yet the fact that Cotes and Newton had something specific in mind with their methodology was lost on the many eighteenth-century natural philosophers claiming to follow in Newton's footsteps because their work contained something mathematical and something experimental.

Perhaps the most important work establishing the Newtonian philosophy as a fashionable subject in the Republic of Letters, especially on the Continent, was the infamous correspondence between Leibniz and Clarke. It took place in 1715–16 and was published in 1717, after Leibniz's death. This controversy, too, started in a context of political turmoil. Perhaps surprisingly for a modern reader, Leibniz's official job at the time was that of court historian of the House of Hanover, whose head became King George I of England after the Hanoverian

Succession of 1714. The correspondence between Leibniz and Clarke was launched by an initial letter from Leibniz to Caroline of Ansbach, the wife of the future King George II, warning Caroline for the religiously dangerous doctrines of "Newton and his followers"; Caroline passed the letter on to Clarke for his opinion, and mediated the following correspondence, in which Clarke has too often been regarded as nothing but Newton's mouthpiece (Bertoloni Meli 2016: 586–96).

Although the controversy went deep into very technical issues such as the question of the existence of absolute time and space, much of its content derived from the different opinions Leibniz and Clarke held on the being of God. In contrast to Clarke's active, intervening, and utterly elevated God whose ways were beyond us, Leibniz's God was primarily one of order and simplicity. According to Leibniz, God had created a completely predetermined a harmonious universe ruled by unbreakable laws of nature that were grounded in the powers of created bodies. Rather than intervening, Leibniz's God was active in the universe only in the sense that he sustained it. Protesting against this view, Clarke argued that Leibniz had put God into a straightjacket and that Leibniz's view smacked of Spinoza's necessitarianism. Leibniz in turn claimed that the God of the Newton and Clarke was not reasonable and powerful enough, comparing him to a watchmaker who had to wind up his clock. It is from this correspondence that we got the historically rather unfortunate view of Newton as having given us the cosmology of a "clockwork universe."

Moving beyond Leibniz and Newton

One element of Leibniz's worldview that challenged the providentialist Newtonians was the activity of matter. In Leibniz's natural philosophy, the notion of "force" took an equally central position as in Newton's but referred to something quite different. According to Leibniz every material body had a certain amount of living force that could only be changed via interactions with other bodies. The total amount of living force in the universe was conserved and this was why God need not intervene in the natural order of things. In Leibniz's dynamics, living force was needed to deal with the phenomena we now explain by recourse to the notions of work or kinetic energy. Many of these phenomena, in particular collisions involving friction and deformation, were not as easily explained in Newton's physics as they were in Leibniz's and this gave rise to what is known as the *vis viva* controversy.

Many mathematicians and natural philosophers, especially in the German speaking lands, took side with Leibniz's on at least some questions in mechanics and philosophy. In philosophy, the most prolific of the authors building on Leibniz was without a doubt Christian Wolff, who saw no need to take into account Newton's foundations for mechanics. Wolff, who was widely seen as an expositor of Leibniz's philosophy but was certainly selective in his reading of Leibniz, aimed for a rationalist system based on a priori principles, especially the principle of sufficient reason which Leibniz had defended with zeal in his correspondence with Clarke. From those principles Wolff attempted to build a secure knowledge. He tried to define the concepts he was using as clearly as possible, a process for which he claimed "mathesis" was of great help. Wolff had an immense impact especially on the teaching of both mathematics and philosophy in Germany. Consequently, in the first half of the century little attention was paid to Newton's mechanics or empirical bend in Germany. This situation changed when disputes broke out at the new Berlin Academy of Science between the adherents of the so-called Newtonian and Leibnizian programs in the 1740s. Leonhard Euler became the main authority in the mathematical sciences in Germany. His adherence to the Newtonian camp, however, was based on the fact that he developed Newton's mechanics, not on his support for Newton's views on God, cosmology, or other philosophical topics, which Euler mostly ignored (see also Chapter 5).

From the mid-century onwards, Immanuel Kant developed an increasingly radical and thorough critique of Wolff's kind of philosophy. In particular, Kant argued that Wolff's approach was useful only on a limited terrain, not including for example, Newton's laws of motion. While Kant no doubt took inspiration from Newton and others who professed to practice Newtonian philosophy, Kant exemplifies how little of the actual Newton was left in the thought of such post-Newtonians. While Newton went to lengths to provide empirical support for the laws of motion, Kant argued that they were synthetic a priori truths and therefore beyond empirical validation. Besides the laws of motion, the main elements of Newton on which Kant built were Newton's gravitational force and absolute space. Kant conceived of gravitation as an immediate action of a body on other bodies across empty space, a view Newton had referred to as inconceivable. Similarly, absolute space, which Newton used as a framework in which to practice physics and needed to refute the Cartesian vortex theory and its theologically unwanted implications, was prominently reconfigured in Kant's *Critique of Pure Reason* (1781) as a form of intuition (Friedman 2019).

Both Euler and Kant used elements from Leibniz and Newton but went far beyond them. They thus ostensibly showed the choice between the two giants of the previous generations was a thing of the past. This illustrates how during the eighteenth century most of the intellectual baggage left by Newton or Leibniz was superseded by new generations of mathematicians and philosophers. The controversy over forces, for example, had started to dissipate when Euler propagated what we now call Newton's second law, and when most functions of Leibniz's living force were taken over by new principles developed in the mechanics of Euler, Pierre-Louis de Maupertuis, and Jean le Rond d'Alembert (see Chapter 5). More generally, due to such practical and technical developments natural philosophy started to move away from metaphysics, particularly in the hands of those who did not see much in the philosophy of Leibniz and Wolff.

Enlightened Mechanics and the Decline of Metaphysics

In "physics," as it became increasingly called in French and Latin, a strand of the Newtonian philosophy arose that distanced itself from systematic and foundational philosophies. This strand was at least rhetorically committed to a search for laws of nature expressed in terms of forces, thus emulating Newton's laws of motion and gravity. Initially, most of the Newtonian philosophers of this brand still argued that the laws of nature in general depended on God's providence. Building on Newton's claim that he would not feign hypotheses, however, they argued that there was not much of a point in looking for the deeper causes of the laws of nature and that we should instead focus finding them by experimental work and describing them mathematically.

One of the clearest instances of this move can be found in Willem Jacob 's Gravesande's *The Mathematical Basics of Physics, Confirmed by Experiment; or, an Introduction to Newtonian Philosophy*, which was published in an impressive array of editions and translations between the 1720s and 1750s. 's Gravesande, a self-proclaimed follower of the Newtonian philosophy who nevertheless propagated Leibniz's concept of force, Leibniz's notation of the calculus, and wrote a metaphysics best qualified as a commentary on Leibniz's *Theodicy*, argued convincingly that neither physics nor metaphysics could prove any of the proposed causes of gravity to be its true cause. Thus, 's Gravesande claimed with Newton, we know only that God had created such laws, not how he did so or by what means they worked (van Besouw 2020).

By deflecting the question of the causes of the laws, the physical debate steered away from inflammatory metaphysical and religious issues. A watered-down but modestly progressive version of Newtonianism emerged that aimed to develop both practically useful knowledge and moral knowledge about God's hand in the universe—without necessarily claiming anything specific about God's relation to that universe (Israel 2006). Under the banner of this Newtonianism, Christian natural philosophers of all kinds united, including particularly many thinkers adhering to atypical forms of Protestantism. And although there was certainly resistance to Newton, and to natural philosophy generally, from within the Catholic Church, this was not as widespread as is sometimes suggested. In fact, there was also much support among particular Catholic scholars. Between 1739 and 1742, two Minim friars based in Rome produced an enormous running commentary on the *Principia*, explaining particularly the mathematics of the *Principia* in great detail. Although further research is required, it seems that their edition became the standard edition in the second half of the century (Guiccardini 2015). One of the minor collaborators of this project, the Jesuit priest Roger Boscovich, played an important role in the vis viva controversy mentioned before. He developed a matter theory compatible with both Leibniz's concept of force and Newton's gravitational physics in his *Philosophiæ naturalis theoria* (1758). When José Celestino Mutis was established as one of the first professors of mathematics in the Spanish Americas in the 1760s, he turned to Newton, 's Gravesande, and Wolff to establish his curriculum, combining a commitment to "the Newtonian philosophy" with overt Catholic proselytization (Orozco-Echeverri and Molina-Betancur 2019).

The somewhat vague providentialist reading of Newton shared by many Protestant and Catholic authors, however, held little persuasive power over those who did not already share most of its assumptions. It is well documented that many of the later readers of for example Clarke's works took a different direction. To name just two, both the polymath and statesman Benjamin Franklin and the empiricist philosopher David Hume digested Clarke's works with enthusiasm, but decided Clarke's opponents actually had the better arguments. More outspoken deists, too, found much to their liking in the *Principia*. Since it implied that the motions of all the heavily bodies could be explained mathematically via the laws of mechanics, it gave them a reason to dispense with an active God. This was of course a move Newton and the vast majority of his earlier British readers would have abhorred.

Arguably the most famous deist interpretation of Newton was Voltaire's. A well-known polemicist, Voltaire had travelled from France to England just in time to attend Newton's funeral. Back in France, he subsequently sparked controversies and brought debates over natural religion, cosmology, and the progressive side of Newton's science into the public sphere. One of Voltaire's methods to attract attention was to dismiss French thoughts and customs and point to the supremacy of the English ways. Comparing Newton to Descartes, Voltaire for example wrote: "very few will presume to compare his [Descartes'] Philosophy in any respect with that of Sir *Isaac Newton*. The former is an Essay, the latter a Master-Piece" (Voltaire 1734: 129–30).

The Newton–Descartes opposition, settled in favor of the former decades before in Britain, had been raised anew at the French Academy of Sciences in 1732 by Maupertuis. Although Voltaire and Maupertuis both believed in a God-creator but not in an active intervening God, they differed much in their social standing and technical capabilities. Whereas Voltaire had a particularly critical view but depended largely on others for his understanding of the technical details, Maupertuis was a man who combined ambition with skill when it came to mathematical natural philosophy. Maupertuis had studied with the Bernoullis and would quickly become a respected member of the Academy. Together, Voltaire and Maupertuis were in an excellent position to reach both specialized and popular audiences and they managed to push the Newton-Descartes opposition into public consciousness (Shank 2008).

A controversy over the shape of the earth took center stage in this push. In the *Principia*, Newton had argued that due to its rotation the earth must be of an oblate shape, being flattened at the poles. This claim was opposed by Jacques Cassini, the director of the Paris Observatory, whose measurements had indicated that the earth had an oblong shape. Since Cassini and other leading members of the Academy of Sciences were attached to Cartesians vortices, the two views were easily cast as a Newtonian and a Cartesian one in the public imagination. To decide which was correct, the Paris Academy prepared two missions, one to Lapland near the North Pole, the other to Quito, in what is now Ecuador, at the equator. The missions were to take geodesic and astronomical measurements, determining the length of an arc of a meridian, but also the length of a pendulum that would beat exactly one second at different places.

These enterprises looked more like voyages of discovery than any other scientific mission before and their results were presented with suitable grandeur. Maupertuis himself set out for Lapland and came back in triumph in 1737. He claimed his measurements proved Newton right—these measurements were compared to those taken in Paris as the Quito mission unsurprisingly faced unexpected setbacks—and concluded that Newton's celestial physics was superior to Descartes'. Even from a mathematical perspective, however, the link to Newton was thin. After consulting with his mentor Johann Bernoulli, Maupertuis decided that Newton's original treatment was incomprehensible and set out to make his own calculations. In the end, the measurements taken by the missions were compared to mathematical models set up by Maupertuis and Cassini themselves.

In any case, the expedition did much to popularize "Newtonianism," a neologism first used by Francesco Algarotti in his hugely popular *Il Newtonianismo per le Dame* (1737). Algarotti was a friend of Voltaire and like him somewhat of a libertine. Detaching the Newtonian philosophy from many of its previous connotations, Algarotti aimed to bring Newton's optical work and philosophical discussion on vision more generally into the "salons" of fashionable cities such as Milan and Paris—the work was immediately translated into French. Popularization of Newton to those yet lacking formal education was also the aim of works such as *The Newtonian System of Philosophy, Adapted to the Capacities of Young Gentlemen and Ladies*, published anonymously in London in 1761, as well as the *Institutions de Physique*, published by the marquise Émilie du Châtelet in 1740 and meant to introduce her son to modern natural philosophy. Du Châtelet, at the time Voltaire's love interest, was arguably the ablest female reader of the *Principia*, and was a genuine peer for the likes of Maupertuis when it came to mathematics and natural philosophy.

In fact, du Châtelet managed to complete a translation of the *Principia* into French—the authoritative French translation—and formed her own theological and methodological views out of bits and pieces from Newton, Leibniz, and Wolff, as Kant would later do (Hutton 2019).

Other ladies were certainly interested in Newton as well, even if very few of them were successful in acquiring access to the means needed to understand

FIGURE 4.1. *Mlle Ferrand thinking about Newton,* Maurice Quentin de La Tour, (1752). Photo courtesy of Helmut Pulte.

works like the *Principia*. Their situation was slightly better in Algarotti's native Italy, where a number of women, most famous among whom was Laura Bassi, holder of a doctorate, was able to participate in the world of natural philosophy (Cavazza 2019).

The subsequent popularity and idolization of Newton in the late eighteenth century is clear in philosophical writings but can easily be grasped through poems and the visual arts, too. Many statues and portraits of Newton appeared. Whenever fashionable people, not least women, commissioned portraits of themselves, they often had something related to Newton laying around. Events such as the return of Halley's Comet in 1758–9, a triumph for Newton's gravitational physics, were celebrated and Newton became an emblem of Enlightenment, particularly of the progressive self-image of the Enlightenment in France. Newton was for instance presented as an exemplar philosopher in the *Encyclopédie* (1751–72) of Diderot and d'Alembert, perceived by many as spearheading the secular Enlightenment.

In the last decades of the century, there was talk about following the Newtonian method in all kinds of sciences. This method usually referred to finding "laws" supposed to govern things as widely diverging as animal behavior and human societies. Apart from being framed as having a Newtonian genealogy, these sciences had virtually nothing to do with Newton and the references to Newtonian origins are best interpret as attempts at historical revisionism. In the high French Enlightenment, Newton came to be seen as someone who had done away with earlier non-constructive scientific approaches and who had opened the way to the natural world. This view was popularized by the political philosopher the Marquis de Condorcet who argued that Voltaire's introduction of Newton in France had caused a "revolution." Like his friend Condorcet, Jacques Turgot credited Newton with bringing light to the darkness which had been thrown over the sciences by the doctrines of Descartes (Feingold 2004: 105–9). Referring to Newton thus meant being modern.

Even those who worked on the sciences addressed by the *Principia* came to disregard what Newton had actually written about such topics as cosmology and causation. Laplace, Lagrange, and the others who would revolutionize gravitational physics and mechanics, certainly considered Newton the herald and starting point of their work. But their physics was a secular one to which revealed religion and metaphysical questions were of little consequence. These authors often presented Newton as if he agreed on that one. Particularly for Laplace, the universe was deterministically ordered and finding out what would happen in the future was a question of applying mathematics to the available empirical data. Since Newton, anything was possible with regard to understanding the natural world. The world had order and man had method. This was how the Enlightenment perceived its debt to Sir Isaac.

PROBABILITY IN THE EIGHTEENTH CENTURY

In 1814, in the second edition of *Théorie analytique des probabilités*, Pierre-Simon Laplace inserted at the beginning of his book a non-mathematical introduction to probability theory. It came to be known as *Essai philosophique sur les probabilités*. In later editions of the *Essai*, Laplace said that it was based on lectures given in 1795 when he taught at the École normale that was established during the French Revolution. The last chapter of the *Essai* contains a history of probability up to the time of the writing of *Théorie analytique*. The topics that Laplace covered in his history include: (1) the early history regarding the analysis of games of chance, particularly through the use of the Binomial Theorem or the expansion of the mathematical expression $(a + b)^n$; (2) the application of probability to the analysis of the length of human life and the application of this analysis to the valuation of life contingent contracts; (3) the analysis of errors made in a set of measurements of some scientific constant, resulting in the justification of the average of the measurements as the best estimate of that constant and the use of least squares as a method of estimation; and (4) the use of Bayes' Theorem to make conjectures about future events based on past experience. In the first chapter of the *Essai*, Laplace also discussed the nature of chance events. This chapter covers the same ground as Laplace while taking into consideration the historical and cultural context of these developments.

The progress of probability in the eighteenth century can be seen as a mixture of the development of theory and the pursuit of application. This was not unique to the eighteenth century; it was already present earlier in the seventeenth century with, for example, the work of Huygens. In 1657, Huygens solved the division of stakes problem in which two players play a series of games, the winner being the first player to win a specified number of games. If play is stopped before either player has won, how should the stakes be divided? Although the problem is set in a gambling scenario, this was essentially a purely theoretical exercise, not a practical gambling problem. Serious gamblers would play out the series of games, or, if forced to quit early, would quickly cast lots for the pot. In fact, this was the solution to a long-standing unsolved problem that appears in Italian mercantile arithmetic books a century and more earlier. On the application side, Huygens and his younger brother Lodewijk corresponded on the topic of how life annuities should be valued. They only got as far as simple life distribution calculations (average future lifetime and median future lifetime) in their correspondence. Gambling and calculations regarding games of chance provided a well-structured model in which calculations for the mathematics of chance could be made.[3]

One historian of probability, Anders Hald, calls the decade 1708–18 the "Great Leap Forward" for probability (Hald 1990). This was the decade in which the theoretical results published by Huygens (Huygens, 1657, *De*

Ratiociniis in Ludo Aleae) were greatly extended and generalized. The major players during this decade were: Jakob Bernoulli, who had died earlier in 1705 but whose major publication in probability appeared posthumously in 1713; Nicolaus Bernouilli, nephew of Jakob; Pierre Rémond de Montmort, an aristocratic virtuoso amateur mathematician; and Abraham De Moivre, a French Huguenot refugee who taught mathematics for a living and did research on the side. Both Bernoullis held university positions, something that Montmort had no need of, but something that De Moivre wished for and never got. They all used games of chance as the model to which they applied their mathematical abilities. In this activity, they had a common motivation of pursuing interesting mathematical problems. Beyond this common motivation, each of them had their own agendas for working on probability problems. Jakob Bernoulli was interested in applying probability to "civil, moral, and economic matters," but died before he could pursue this interest very far. Nicolaus Bernoulli took up his uncle's fallen probability banner and wrote his doctoral thesis on these very issues. For Montmort, mathematics was an interesting pastime; the parallel for the rich today is performing at a high level at expensive sports activities or becoming an art connoisseur and collector. What initially drove De Moivre were questions about probabilities in games of chance asked of him by aristocratic and other wealthy patrons. These questions motivated him to use and then to develop new mathematical tools that were more widely applicable than the problems he was initially given. He was also driven to dig deeper into probability theory because of accusations of plagiarism by Montmort.

Through their mathematical researches, these mathematicians believed they had in a sense tamed chance (cf. Hacking 1990). Fortuna, the fickle goddess of fate and fortune had lost her power over chance events. There was now a mathematical structure that shaped the random outcomes of games of chance. Just as order in the vast parts of the physical world had been illustrated through Newton's *Principia Mathematica*, a sense of order in the previously perceived chaos of chance could be relied upon through mathematics. This is made explicit in two allegorical engravings, one appearing in Montmort's 1708 *Essay d'analyse sur les jeux de hazard* and the other in De Moivre's 1718 *Doctrine of Chances*.

On the left side of Montmort's engraving, Montmort is showing a piece of paper with some mathematical symbols on it to Minerva, the Goddess of Wisdom. Montmort is standing in a dominant position over Minerva showing that his work in chance has superseded what the ancients knew of the subject. Mercury, the messenger of the gods and the god who presided over dice games, is seated on the right of the engraving. He is on the same level as the players at the table. This shows that the gods no longer have control over the outcomes of games of chance. Instead, it is Montmort through his mathematics who has given laws to chance. It is the monkey who identifies Montmort as the

FIGURE 4.2. Engraving in Montmort's *Essay d'analyse sur les jeux de hazard* (1708). Photo courtesy of Stephen Stigler.

mathematician who has done this. On the chess board, the monkey is pointing to a knight's move from an initial bishop's position. Montmort had been a canon in the church in France as Pierre Rémond, but left the position to get married and establish himself as a landowner, Pierre Rémond, Sieur de Montmort.

Montmort was convinced that Abraham De Moivre had plagiarized his work. What De Moivre, the better mathematician, had done was to provide more general and simpler proofs to the probability problems than Montmort. This is illustrated in De Moivre's allegorical engraving that references one of the major probability problems of the era, the problem of the duration of play. It is related to another problem called the gambler's ruin problem. The scenario for both of these problems is that two players, each with different amounts of money at the commencement of play, carry out a series of bets against each other until one of the players runs out of money. The gambler's ruin problem is, given a player's probability of winning each bet in the series and the player's and the opponent's initial capital, what is the probability that the player exhausts his capital before his opponent does (i.e., is ruined)? The duration of play problem is related. With the same setup, what is the probability that one of the players is ruined within a given number of bets in the series? These are difficult mathematical problems, whose only practical application that I can see is that it is very profitable to run a well-capitalized casino. The casino's capital being much greater than that of an individual player, it is almost certain that the player will exhaust his funds before the casino is ruined. De Moivre solved the duration of play problem using a geometrical argument. It resulted in a very quick and easy way to get a numerical solution to the problem. Earlier solutions were long, cumbersome, and difficult to evaluate numerically. In De Moivre's engraving, on the left the Goddess of Wisdom is instructing Fortuna, the Goddess of Chance about De Moivre's geometrical argument. Fortuna is holding a diagram that is the essence of De Moivre's solution. De Moivre, who does not have the effrontery to talk to the gods directly, is instructing other mortals about chance. Underneath the cornucopia beneath Fortuna's wheel is some mathematical writing, indicating that some earlier writing in probability has borne fruit. This is probably a reference to Huygens and does not include Montmort's work. The two boys in the middle are reading De Moivre's *Doctrine of Chances*. They have thrown aside their dice and cards, as well as a chess board. The chess board, a reference to Montmort's engraving, is not a square board indicating that Montmort's work is incomplete.

Over the course of the eighteenth century there was an evolution in thinking about the nature of a chance event. It is useful to start much earlier, in 1619, with Thomas Gataker, an English Puritan divine who gave a lot of thought to chance events when considering whether or not it was appropriate to play games of chance. Gataker was working within a deterministic world where everything was subject to divine providence or the controlling hand of God. To Gataker, a chance event for an individual was something whose outcome was the result of natural causes (God's providence) but at the same time unpredictable to that individual. Many of Gataker's Puritan colleagues thought that God intervened directly to determine the throw of the dice (God's special

FIGURE 4.3. Engraving in De Moivre's *Doctrine of Chances* (1718). Photo courtesy of Stephen Stigler.

providence). Gataker further clarified his meaning by saying, "that neither skill or counsell hath any hand in the directing of those causes in the producing of that effect, his fore-cast can determine what the effect will be in particular but by meere conjecture onely" (1619: 12). One hundred years later, the German philosopher Christian Wolff had a more universal definition of a chance event, rather than one that applies to a particular individual. Also working within a deterministic world, he saw a chance event as coming from a sequence of innumerable causes working before and during the occurrence of the event. The causal sequence is complex enough that it cannot be unraveled (Wolff 1720: 313). By the end of the eighteenth century, Laplace, in his *Essai philosophique sur les probabilités*, had completely removed God from any considerations regarding chance. Similar to Wolff, Laplace held that all events stem from some physical cause or causes. It is ignorance of the causes that leads to probability statements. He concluded, "that it is not surprising that with the same data two persons arrive at different results, especially in very complicated questions" (Laplace 1814).

During the decade of the Great Leap Forward and for the next couple of decades, there were several enduring mathematical results that were obtained related to what are now called Bernoulli trials. These trials are events with only two outcomes, often dubbed "success" and "failure." The duration of play and gambler's ruin problems are two such types of problems. Both problems are based on winning (success) and losing (failure) in a series of bets. Other major problems solved were what are now known as the weak law of large numbers, published by Jakob Bernoulli in 1713, and the normal approximation to the binomial, published by Abraham De Moivre in 1733. Both results apply the count of the number of successes in a fixed number, but large in size, of Bernoulli trials with a constant probability of success on each trial. Bernoulli showed that the proportion of successes observed in the trials will be approximately the probability of success for each trial. De Moivre showed that probabilities related to the total number of successes can be accurately approximated through the normal probability distribution (or a bell-shaped curve). More or less ignored in his lifetime, De Moivre also showed that when the number of trials is large and the probability of success on each trial is small, probabilities related to the total number of successes can be accurately approximated through what is called the Poisson probability distribution.

Montmort's *Essay d'analyse sur les jeux de hazard* and De Moivre's *Doctrine of Chances* contain descriptions of various games and the probabilities of winning them. And yet this literature had a minimal effect on gambling manuals and the literature of gambling. The various publications throughout the eighteenth century of Edmond Hoyle (from whom we derive the phrase "according to Hoyle"), and of his intellectual descendants in English, and of the French manual by various authors on games, *Académie universelle des jeux*, were devoted mainly to describing the rules of games in order to combat cheating and to reduce arguments. The one concession to probability that crept into these publications was advice on the strategy of play in various situations, advice that was devoid of any numerical calculation on the part of the reader. For people like Montmort and De Moivre, the attraction of games of chance was that in their complexity; they provided interesting and difficult mathematical problems to solve. Later in the eighteenth century, mathematicians found other areas in which to apply and build on the general tools that had been found earlier. For the player of games of chance, the main issue was how not to be cheated in the game. This issue remains with us today.

With respect to probability theory applied to the chance of human life and the valuation of life contingent contracts, several early attempts were made on both sides of the English Channel to crack this nut. The British were the most successful in this endeavor. One of two nutcrackers was needed—either a numerical life table showing the chances of death at each age or a mathematical model that would do the same thing. Based on some data available to him, in 1671 Johann de Witt formulated a simple mathematical model for human survival and used it to calculate the value of life annuities used as a financial tool by the Dutch Republic. He submitted his report, in Dutch, to the National Assembly of the republic. Later, in 1709, in his doctoral dissertation, Nicolaus Bernoulli presented a simple mathematical model for human mortality that assumed the chances of survival decreased linearly with age. Neither of these efforts had much impact. De Witt's report was not disseminated beyond the National Assembly and he died amid political turmoil a year after his report was submitted. Bernoulli's work was written in Latin and also was not widely disseminated beyond a small academic audience. Leaves fell in the forest and there was no one there to hear them make a sound. In 1693, Edmond Halley was the first to construct a complete life table by age based on data collected in the city of Breslau (now Wrocław). There was much interest in Halley's work, but at the time it had little practical impact. The arithmetical calculations necessary to obtain the value of a life annuity were lengthy and onerous. Only a very few spent the effort to make the necessary calculations.

In 1725, Abraham De Moivre also assumed a mathematical model for human mortality, which, like Nicolaus Bernoulli's, had the chances of survival decrease linearly with age. De Moivre was unaware of Bernoulli's work. Published in English, De Moivre's work on the valuation of life annuities had enormous impact. With the model based on Halley's life table (the only one available at the time), De Moivre was able to accurately approximate annuity values without having to carry out tedious and lengthy arithmetical calculations. Today, De Moivre and Halley are seen as early the pioneers of actuarial science, or insurance mathematics. However, De Moivre and other early English actuarial pioneers, including probably Halley, were not motivated by insurance problems.

Rather, their motivation came from valuing life contingent contracts related to land and property for their aristocratic and landed patrons. Most of the annuity problems solved by De Moivre in 1725 and later have direct application to property and very little connection to insurance.

One example of this is a 1777 contract on which the English philosopher and mathematician, Richard Price provided input as a consultant. A young man in England inherited an estate on which there were conditions. If he died without having any male children, the estate would go to a cousin, again under some conditions. If that cousin also died without having any male children, the estate would go to another cousin. The problem was that the young man wanted to get married and needed to have a marriage settlement in order to do so. This involved setting aside part of the estate such that if he died before his wife, the rents from the part that was set aside would be used to support his widow during her lifetime. He could not do this outright, as it infringed on the potential property rights of his cousins. The solution was to set up a fund that would cover the expected losses his cousins might incur should the young man die before his intended wife and also die without a male heir. Using De Moivre's survival model, Price calculated the size of the fund that was needed and how the fund should be divided between the two cousins in the event that either of them inherited the estate.

The stability of the proportion of successes observed in a series of Bernoulli trials becoming approximately the probability of success for each trial, established by Jakob Bernoulli, found its fullest flowering in the last quarter of the eighteenth century. It was through the use of life tables by life insurance companies, initially established in Great Britain. The first to offer life insurance priced by sound actuarial theory was the Equitable Life Assurance Society in 1762. Probabilities of death in the form of the number of deaths (the successes in this case) divided by the number at risk in the population (the total number of trials), were used to calculate the premiums to be charged for life insurance policies. Richard Price became an actuarial consultant to Equitable Life and helped put the company on a sound footing.

There was a theological motivation to Price's interest in probability and insurance; it might be called the social gospel. Price and others saw the development of theory for the valuation of annuities as a means to providing old age security for those whom he called the industrious poor. He was successful in encouraging cooperative insurance (or friendly) societies, but was not successful at national schemes funded by the government because of opposition from landed interests. Price did have a great impact on the early insurance industry in the private sector, both within and beyond England. For example, based on Price's principles, a French company, Compagnie royale d'assurances was founded in 1787. It did not survive the French Revolution. Life insurance in France remained in abeyance until after the Napoleonic Wars.

Halley's work on life tables stimulated others in various European countries—the Dutch Republic (Nicolas Struyck and Willem Kersseboom), France (Antoine Deparcieux), Prussia (Johann Peter Süssmilch) and Sweden (Pehr Wargentin)—to collect and use data for the construction of life tables. Some went beyond Halley's observations. Struyck, for example, showed that there were differences in male and female mortality. Later, Deparcieux and Wargentin separately showed that females lived longer than males and that married women lived longer than single women. Beyond demographic observation, Süssmilch was interested in making an argument for the existence of God based on the demographic regularity of human life. Christian Wolff wrote the preface to Süssmilch's book, *Die göttliche Ordnung in den Veränderungen des menschlichen Geschlechts* published in 1741. Wolff was also interested in these theological questions as they were connected to the nature of a chance event. Struyk, Kersseboom, and Deparcieux all made life annuity calculations based on their life tables. Their impacts were much less than De Moivre's. Unlike England, neither the Dutch Republic nor France had laws of entail, which determine inheritances of property. As seen from Richard Price's consultation in 1777, entailment of property was the major stimulus for De Moivre's work on annuities.

After about 1750, the major incentives for the development of probability on the Continent came from scientific questions, Newtonian celestial mechanics in particular, and the patronage behind it. With respect to patronage, in France, for example, through the Académie royale des sciences de Paris, there was state support for scientists. The Royal Society in London was not given state funds to make this support possible. Mathematicians like De Moivre had to earn their living while carrying out their mathematical research on the side. The incentives that are available often control the direction of the research that is carried out. De Moivre's incentive to work on life annuities, for example, was economic rather than scientific.

The major work in probability motivated by celestial mechanics originated from a simple observation in the previous centuries that when measuring a scientific constant in two or more different ways you can get slightly different results. This is called measurement error. The basic question now arises. With several measurements of the same thing, what single representative value should be used to value the scientific constant? By the mid-eighteenth century, scientists had settled on the simple average of the measurements, but could not say why this was a good representative value. The starting point was to assume a model for measurement error. In celestial mechanics, deterministic models had been used to model the heavens. And the models worked well. Measurement error was different. Direction and size of the error was unpredictable. Similar to the length of human life, mathematicians assumed a model for measurement error that was stochastic in nature rather than deterministic.

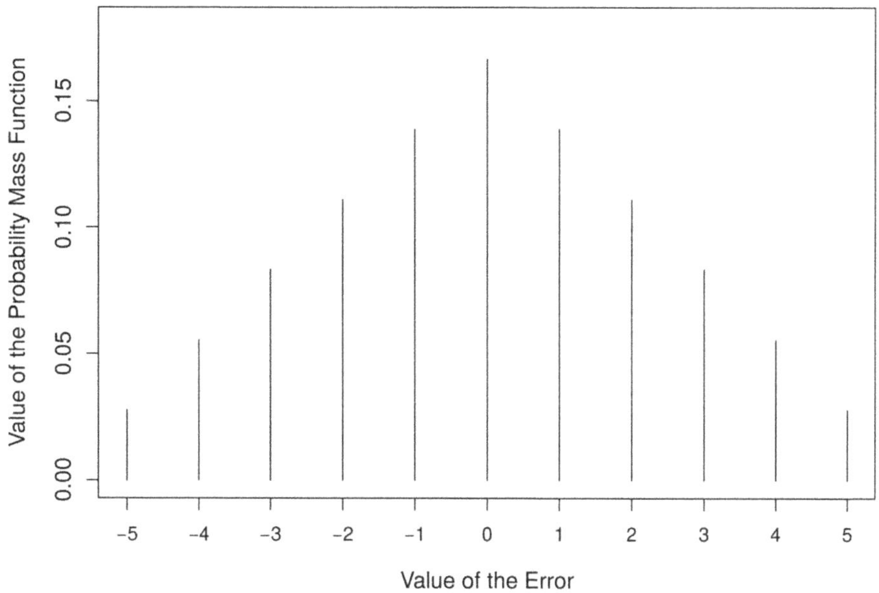

FIGURE 4.4. Simpson's Error Distribution. Image courtesy of David Bellhouse.

Models for measurement error in the eighteenth century built on ideas from Galileo's 1632 *Dialogue Concerning the Two Chief World Systems – Ptolemaic & Copernican*. Galileo commented on two aspects of the distribution of measurement errors. These are in his words (in translation): (1) measurements "are equally prone to err in one direction and the other" (p. 291) and (2) "it is plausible that the observers are more likely to have erred little than much" (p. 308).

In 1755, Thomas Simpson in England was the first to attempt to derive a distribution for measurement errors and to attempt to put the use of the sample mean as an estimator on a firm theoretical foundation. His distribution was a simple one; he considered an astronomical instrument which measures in seconds with a maximum error of ±5 seconds. To justify the use of the simple average, Simpson calculated the probability that the error in a single observation exceeds a certain amount (1, 2, or 3 seconds) and compared it to the probability that the average of 6 errors exceeds the same amounts. The probabilities related to the averages were always smaller than for single observations. Further, the probabilities for the averages decreased much faster than those for single observations as the number of seconds increased.

Others tried their hands at error distributions and at a justification of the average as a good representative value. In 1774 and 1777, Joseph-Louis Lagrange and Daniel Bernoulli respectively published results on some

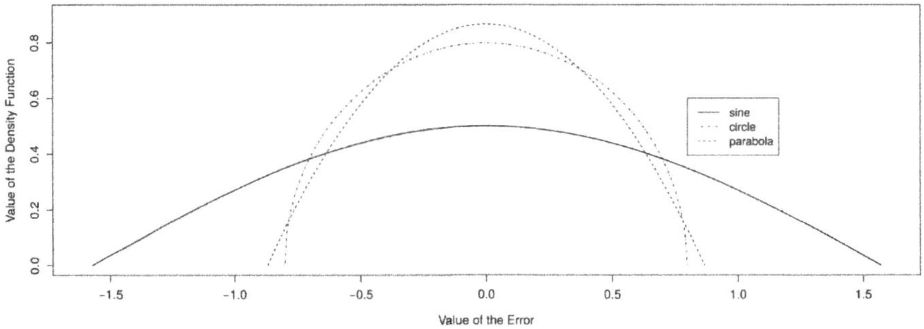

FIGURE 4.5. Lagrange's and Bernoulli's Error Distributions. Image courtesy of David Bellhouse.

continuous error distributions rather than the discrete one that Simpson had used. Lagrange tried using a parabola and a sine curve as error distributions, while Bernoulli used a semi-circle. One problem with them is that the bounds on the size of the errors are finite so that the maximum size of the errors must be known in advance. In 1774, Laplace solved this problem with a curve that decreases exponentially away from 0. All these models shared a common problem. The distribution of the average of the observations was difficult to find or difficult to describe in a simple fashion. The whole problem remained in abeyance until 1809, when Carl Friedrich Gauss justified the use of the normal (or bell-shaped curve, aka the Gaussian distribution) as an error distribution. This gave a simple distribution for the average, or indeed for any linear combination of the observations.

The more general data analytic problem is to take several measurements (that have measurement error) of related objects that have a relationship to some values of interest that are unknown. The relationship is set out in a linear equation. How should the data be combined to obtain estimates of the unknown values? If the number of observations made on the sets of measurements is equal to the number of unknown values then a unique solution can be obtained for the unknowns. Because of measurement error, it is usual to take more observations than the number of unknowns so that the number of equations is greater than the number of unknowns. Many mathematicians tried unsuccessfully to solve this problem by combining the equations in some way to reduce the number of equations to the number of unknowns.

The successful approach involves finding the difference between what was observed and what the linear equation would predict for the observed value. Call this difference *d*. Those who used this approach were motivated by trying to measure the earth, in particular the length of a meridian arc. To verify Isaac Newton's claim that the shape of the earth is ellipsoidal, the Jesuit mathematician and astronomer Roger Boscovich used arc lengths on the meridian passing near

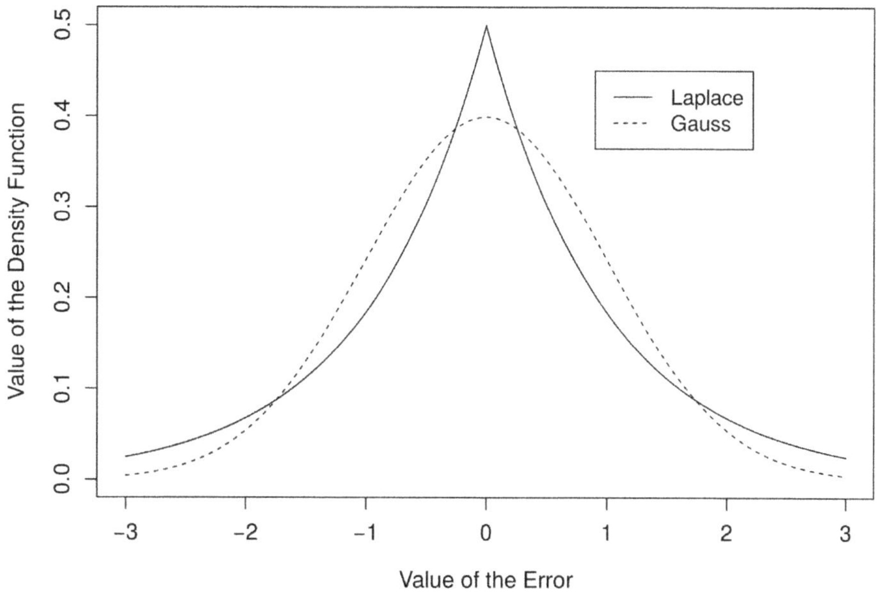

FIGURE 4.6. Laplacian's and Gaussian's Error Distributions. Image courtesy of David Bellhouse.

Rome to estimate the ellipticity of the earth in 1755. Motivated by the French revolutionary government's desire to have a standard set of weights and measurements (the metric system), in 1805 the French mathematician Adrien Marie Legendre used the meridian arc lengths passing through five locations in France to obtain the length of a meter. The meter had been set at one 10-millionth of a quadrant of the meridian. Both Boscovich and Legendre had five linear equations with two unknowns in the equations for Boscovich and three for Legendre. Boscovich minimized the sum of the absolute values of d and Legendre took what is now the standard approach today of minimizing the sum of squares of the values of d. Without the use of a computer, Legendre's method of least squares is much easier to carry out than Boscovich's method of least absolute deviations.

Shortly after Legendre's 1805 work, Laplace was able to tie the development of the method of least squares to results on error distributions. As already noted, Gauss justified the use of the normal, or Gaussian distribution, as an error distribution. Once Laplace had seen Gauss' work, he saw the connection between error distributions and the estimation of the unknowns in linear equations. In 1810, Laplace showed that all estimates that are linear functions of the observations that have normally distributed measurement error are also normally distributed. Among these estimates, the least squares estimate has the smallest expected error. He also obtained a version of the central limit theorem for the sample average of the observations: for a "large" number of observations, the sample average has an approximate normal distribution. Laplace's results laid the groundwork for modern statistical analysis.

Laplace had enormous influence on modern statistical analysis in another way. This was through the use of Bayes' Theorem, published by Thomas Bayes in 1763, but better expressed in a more general context by Laplace in 1774. While Jakob Bernoulli's and De Moivre's results related to the number of successes observed in a fixed Bernoulli trials showed regular structure for what happens in future events (the successes are *to be* observed—probability is speculation about what happens in the future), Bayes' and Laplace's results deal with how to use the experience of the past (the successes that have been observed in the past) to predict the occurrence of a success in a future experiment. The approach that Bayes and Laplace used may be expressed generally as: the posterior probability (the probability *after* considering the data collected) is proportional to the prior probability (the probability expressed *before* the data are collected) times the likelihood of the data obtained. Both Bayes and Laplace used uniform prior probabilities (all values are equally likely to occur) to express their lack of knowledge about unknowns such as p. This came to be known as the principle of insufficient reason.

Bayes' work was, for the most part, ignored for the rest of the eighteenth century due to the difficulty in calculating the integral to obtain the probability. Without knowing Bayes' result, Laplace independently solved Bayes' problem and was able to obtain a good approximation to the required integral that made the result practicable. Laplace's expression of Bayes' Theorem is what appears in most modern elementary probability and statistics textbooks. Laplace's original work in probability, and further developments in the 1780s and beyond, combined with the application of the principle of insufficient reason had an enormous impact on probability and statistics throughout the nineteenth century and well into the twentieth century. Bayesian methods are standard approaches to complex data analyses today.

As a brief aside, it was Richard Price who presented Bayes' mathematical results to the Royal Society in 1763. In the resulting published paper, Price provided Bayes' theoretical results as well as his own applications to Bayes' Theorem. Price saw Bayes' result as a new method of inductive inference. He also saw the theorem as a proof of the existence of God. He argued that through probability theory there are fixed laws through which things happen, and this supports the argument of final causes that had been used in the proof of the existence of God.

Over the eighteenth century, there was another approach put forward to handle inductive inference, called "moral certainty" by Georges-Louis Leclerc, Comte de Buffon. The approach is related to proof by contradiction. Suppose we have two statements; call them A and B. The logic goes as follows: (1) "A implies B" is a true statement; (2) "B" is not a true statement; (3) therefore, "A" is not a true statement. Inductive inference arises when, instead of "B" being a false statement, it is unlikely to be true. Applying Bayes' Theorem would give you the likelihood that "A" is not true. Instead, Buffon's approach was to determine the probability that "B" is true. If that probability is sufficiently small, then Buffon would decide that it was morally certain that "B" was not true and proceed to conclude that "A" is not true. The key phrase for the probability is "sufficiently small." Buffon thought the probability should be less than 1/10,000. Daniel Bernoulli thought Buffon's number should be 1/100,000. The approach Buffon and Bernoulli took is the basis of modern statistical hypothesis testing.

CHAPTER FIVE

Describing and Understanding the World

From Newton's Principia to Laplace's Traité de mécanique

HELMUT PULTE

Like mathematics itself, "describing and understanding the world" is a *cultural* activity: Both are conceptual and constructive works, embedded in social and historical contexts. Mathematics can be used as an aid or tool in order to do these works, or mathematics may be an essential and integral constituent of these works. In what follows, I will argue that the second meaning is appropriate for the period in question, and that no adequate understanding of the formations and transformations of concepts, laws, and theories meant to describe and understand the world can be gained without it.

The eighteenth century is often described as the "Age of Enlightenment" or the "Age of Reason." According to Jean le Rond d'Alembert, himself an eminent mathematician and philosopher, the beginnings of this age go back to the seventeenth century, with Francis Bacon, John Locke and Isaac Newton as its first ancestors. This chapter will start with Newton's *Principia* from 1687, and does so *not* because I share the view held by eminent historians like Thomas S. Kuhn or I. Bernard Cohen that this work marked an unparalleled and *revolutionary* break with the past, but in contrast that in a certain sense it was the completion of a longer development—comparable to the place of Euclid's *Elements* in the history of ancient geometry, and with remarkable structural similarities to this work.

The *Principia* is a work on "rational mechanics," to use Newton's own label, or on theoretical mechanics, as we would label it today. The larger part of this chapter will focus on this discipline. From a modern point of view, such a focus is anything but self-evident and requires further explanation: If, in order to roughly structure the scientific scenery of the period in question, we adopt Thomas Kuhn's well-known distinction of "classical sciences" and "Baconian sciences" for the moment, we have to deal—next to mathematics (arithmetic, geometry)—with astronomy, optics, statics, and harmonics on the one hand, and experimental sciences, having to do with phenomena like heat, electricity, magnetism, etc. on the other hand (Kuhn 1977: 31–65). The sciences of the first kind were already practiced in antiquity and were at least partly conceptualized in a mathematical manner. The sciences of the second kind emerged in (early) modern times and were—based on different methodological and epistemological premises—mathematized only at an advanced stage of their development. Mechanics, however, understood here as science of motion, has no place in this basic distinction: it was—for philosophical reasons—understood as an art rather than a science in antiquity, rudimentarily developed as a mathematical science in the Middle Ages and played a certain role in practical contexts in early modern period, but then gained enormous importance from the seventeenth century onwards.

Why is that? Next to tackling practical problems, such as determining the trajectories of cannonballs, the answer lies in the revival of Greek atomism in the Renaissance and its further development into a new corpuscular philosophy by people like Galileo Galilei, Thomas Hobbes, Pierre Gassendi, Robert Boyle, and later René Descartes. The fundamental idea of advanced corpuscular philosophy was that matter is constituted by small corpuscles or atoms, of which the basic or constitutive properties like shape, size, mass, location in space and time and thereby motion are quantifiable, while other properties like color, taste, heat, sound, etc. are secondary or derivative, i.e., are either mere subjective impressions of the observer or (often: and) can be traced back to the primary properties (or "qualities"). Based on such a metaphysically imbued theory of matter, often flanked by philosophical premises of Neoplatonism and Neo-Pythagoreanism, mechanics as the science of motion became fundamental for modern science in general. If you understand how particles move, how motion arises and how it is changed in the interaction with other particles, you can "describe and understand the world." Phenomena that still need to be explained, such as phenomena of heat, electricity, magnetism, chemistry, or even of living bodies, can then be traced back to mechanics at the end of the day. And this means that, according to the program of corpuscular philosophy, the world can be described and understood *mathematically* in a very specific sense. As the basic concept of mechanics are quantifiable, and as their relations in forms of "laws of motion" are *thereby* (an anything but trivial conclusion) expected to be mathematical laws, the science of mechanics was understood by most representatives of this program as a *genuine mathematical* science. In this sense, we do not have to deal with mechanics as a discipline, where mathematics is "applied" to nature, but rather as a science that has to do essentially with mathematical entities: space, time, velocity, mass etc. Moreover, mechanics is thus understood as a discipline that bridges the gap between "classical science" and "Baconian sciences" in Kuhn's sense. In rational mechanics itself, experiments play only a secondary role. However, observation and experiments in areas like heat, electricity, etc. are viewed as important to find out regularities ("candidates for laws") which must be, in a second step, mathematically conceptualized and ideally, in a third step, traced back to the fundamental laws of mechanics. This is the program of "classical mechanism," as I label it, and at the same time the outline of an explanation why mechanics takes center stage in this chapter. In the following main section, I will try to give a structural overview of the development of rational mechanics in the Age of Enlightenment, including its later problems with the general claim to be a genuine *mathematical* science. However, the third section will broaden the view to other fields, such as astronomy and other areas.

"THERE IS ONLY ONE SYSTEM OF THE WORLD TO BE DISCOVERED": RATIONAL MECHANICS FROM NEWTON AND LEIBNIZ TO EULER AND LAGRANGE

Newton's Principia *(1687)*

Isaac Newton's *Philosophiae naturalis principia mathematica* from 1687, sometimes described as the "bible" of classical physics, is, without any doubt, one of the most impressive achievements of the so-called Scientific Revolutions and indeed of a mathematical representation of natural phenomena in general. Though this view is widely shared in history and philosophy of science, there is less agreement how it can be substantiated. Is it Newton's foundation of mechanics in the first book of the *Principia* "On the Motion of Bodies," based on his three famous laws of motion, that makes this work unique? Most interpreters of the book have answered this question in the affirmative. Hardly anyone attested the second book, dealing with the movement of bodies in resisting media, a revolutionary character. However, the third book on "The System of the World," described by Newton—in contrast to the first two "mathematical" parts—as the "philosophical" part of his work, including his universal law of gravitation, is another and often mentioned reason for giving Newton's *Principia* an exceptional status. I agree to this but reject the view that the foundation laid down in the first book is similar in character: Newton was *not* the one who founded classical theoretical mechanics by establishing for the first time a set of first laws of motion. Contrary to popular wisdom, "Newtonian mechanics" and "classical mechanics" are by no means identical.

Newton himself made no claim to priority regarding his laws of motion. Obviously, he knew that in this respect the *Principia* was rather a synthetic work, building on Galilei Galileo's *Discorsi*, René Descartes' *Principia philosophiae*, Huygens' *Horologium oscillatorum* and other achievements aimed at revealing first principles of motion. Despite these precursors, the *Principia* occupy a special place: one can compare its position in the history of mechanics with that of Euclid's *Elements* in the history of geometry. Both works integrate older scientific knowledge und bring it into a certain logical structure without contributing any particularly original achievements of their own. Nor was the older knowledge the result of a steady development, starting from common conceptual and methodological premises. In this sense, the *Principa* were not "revolutionary" as Kuhn understood the term, *because* they could not build on a tradition of "normal science" (cf. Kuhn 1996).

The *Principia* and Euclid's *Elements* are also similar in a different sense. Obviously, Newton followed Euclid's axiomatic-deductive structure, starting from mathematical definitions of basic concepts (like quantity of matter and motion, inherent and impressed force, etc.), explications of absolute and relative space and time, and *axiomata sive leges motus* as first principles of motion on which, in his opinion, the whole logical structure of the *Principia* is based. This structure was not changed despite other important amendments and alterations, in the later editions of the book (2nd 1713, 3rd 1726). In the (now standard) translation of the third edition, the principles of motion are given as:

> *Law 1*: Every body perseveres in its state of being at rest or of moving uniformly straight forward, except insofar as it is compelled to change its state by forces impressed.
>
> *Law 2*: A change in motion is proportional to the motive force impressed and takes place along the straight line in which that force is impressed.
>
> *Law 3*: To every action there is always an opposite and equal reaction; in other words, the action of two bodies upon each other are always equal and always opposite in direction. (Newton [1726] 1999: 416–17)

The first law or "law of inertia" was well known long before Newton (Gabbey 1980), as were various forms of the third law (Russell 1975), often abbreviated as "action equals reaction." The reader will note that the second law is not expressed in terms of acceleration a ($F = m\, a$, as it is known today and ascribed to Newton). The "fully modernized" form of this law, as it can be found in today's textbooks, is this (directed magnitudes underlined):

$$\underline{F} = m\,\underline{a} = (m\frac{d^2x}{dt^2}, m\frac{d^2y}{dt^2}, m\frac{d^2z}{dt^2}) \qquad (1)$$

However, in Newton's words, the second law expresses a proportionality of the force impressed to a body and its "change in motion," what we today would describe as change of its impulse $\underline{p} = m\,\underline{v}$. This means that a formal representation of the *original* second law, still avoided by Newton, looks like this:

$$\underline{F} \sim \Delta\, p = \Delta\,(m\,\underline{v}) \qquad (2)$$

with $\underline{F}$ being the directive "motive force impressed." It is illuminating for understanding how mathematics is used in the *Principia* to note the reasons why we do not find the "modern" form *(1)* in the *Principia*, but the original form *(2)*: For one, although Newton was one of the discoverers of the infinitesimal calculus, he avoided writing the *Principia* in its language, but chose the traditional, "synthetic" language of geometry in order not to add to the burden of understanding the book. On the other hand, Newton's theory of matter included the existence of perfectly *hard* atoms. As he claimed universal generality for his first principles, the second law should be applicable to the collision of hard bodies, too. However, as such a collision would happen without deformation of the bodies and therefore in an infinitesimal period of time ($\Delta t \approx 0$), form *(1)* would yield potentially an infinite large magnitude of force, which is impossible for Newton, because a mathematical representation of nature must make do with *determined* magnitudes. This exemplifies how philosophical convictions or, to be more precise, *scientific metaphysics* influences and, indeed, largely *shapes* the mathematical approaches to rational mechanics in Newton's time. We will see how and why this relation changes later in the eighteenth century.

How, then, does Newton use his more or less "traditional" laws of motion to reach the important new results of the *Principia*? In particular, how does he use the second law in a sense which is close to its later meaning? Let us consider at least one example to illustrate the more general answer to these questions: Theorem 1 of Section 2 of the first book. It reads: "The areas which bodies made to move in orbits describe by radii drawn to an unmoving centre of forces lie in unmoving planes and are proportional to the times" (Newton [1726] 1999: 444).

With Theorem I, Newton presents his formulation of Kepler's laws of areas. A planet's trajectory is approximated by a polygon, where the corner points A, B, C . . . are defined by small and equidistant periods of time. In the first period, the planet, attracted by a force directed to the sun S, moves from A to B, in the second period from B to C, etc. Without the centripetal force the body would, if we take for example position B, not move to point C, but—according to the

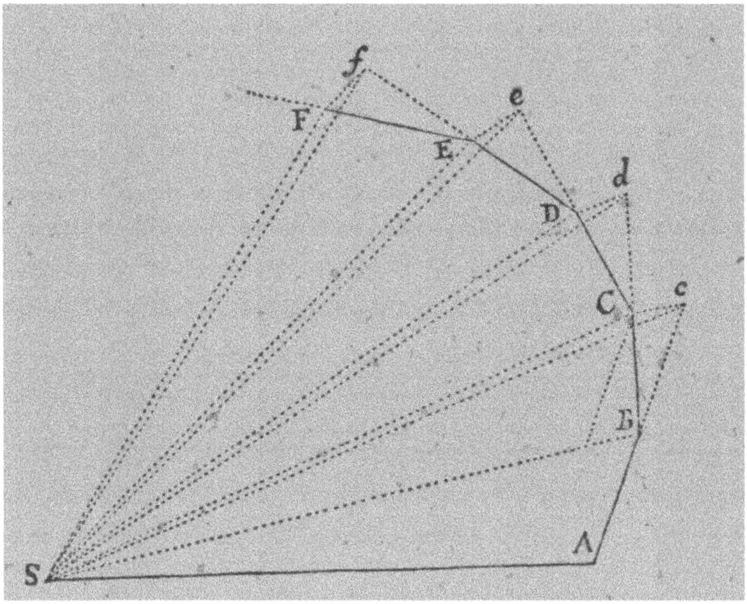

FIGURE 5.1. Polygonal approximation of a planet's motion in order to demonstrate Kepler's area law from Newton's *Principia*. Photo courtesy of Universitätsbibliothek Basel, UBH AN VI 17b, Public Domain.

law of inertia—tangential to point c (with Bc = AB). Elementary geometry shows that the areas of the triangles SAB and SBc are equal. In fact, the central force *F* inserts, according to the second law, an "impulse" $\Delta\ (m\ \underline{v})$ on the body that causes the deviation from c to C, with cC parallel to BS. In order to determine C, and to show that the areas of the triangle SBC and SBc are equal (henceforth, also those of SBC and SAB), Newton makes use of the rule of parallelogram, already demonstrated earlier as a corollary to the first and second law of motion. The equality of areas holds not only for SAB and SBC, but also for SBC and SCD, etc. This means that the body moves in a fixed plane, defined by SAB and—by summing up—the areas of different parts of the polygon, defined by corner points with equal distances of time, is the same. Newton's last step of his demonstration reads: "Now let the number of triangles be increased and their width decreased indefinitely, and their ultimate perimeter ADF will [. . .] be a curved line; and thus the centripetal force by which the body is continually drawn back from the tangent of this curve will act uninterruptedly, while any areas described, SADS and SAFS, which are always proportional to the times of description, will be proportional to those times in this case. Q.E.D." (Newton [1726] 1999: 445).

This specimen from Newton's *Principia* and its further implications are very remarkable for more than one reason. First, Theorem 1 (and the subsequent, somewhat "inverse" Theorem 2) expresses the important mathematical insight that motions governed by central forces follow Kepler's area law (and vice versa). Second, the theorem and its demonstration illustrate how Newton geometrically represents continuously acting forces and the time of action. Thirdly, the approximation process involved in the demonstration exemplifies how Newton "vested" his methods of fluxions, i.e., his version of the infinitesimal calculus, in "geometrical clothes," though some intimations to his calculus can be found in the *Principia*, when he justifies his use of "first and ultimate ratios" (Newton [1726] 1999: 433, 441–2). Moreover, fourthly, we get an impression how the modern form *(1)* of the second law could be read in Newton's use of his original form *(2)* in the case of continuously acting forces $\underline{F}$. If this force is meant to be proportional to $\Delta\ (m\ \underline{v})$ for small and equidistant periods of time Δt, we can interpret *(2)* as expressing the impulse $\underline{F}\ \Delta t \sim \Delta\ (m\ \underline{v})$ for a "very small" period of time. If we assume the mass m to be constant (which in most mechanical processes is the case), and if we ignore details concerning physical units here, equation *(1)* is gained from this proportionality by division of Δt and by taking the limit $\Delta t \rightarrow 0$. However, this would be a simplifying interpretation by hindsight, ignoring other contributions to the modern form *(1)* and later developments.

Finally, Theorem I and its demonstration shed light on how Newton understood the mathematical character of the *Principia* and the general role of mathematics for "describing and understanding the world." Without doubt, he considered Kepler's law of areas as a mathematically demonstrated truth, deduced from his laws of motion, geometrical laws and antecedent conditions (like "bodies made to move in orbits described by radii drawn to an unmoving centre of forces"). Is Kepler's law, then, a special empirical law of nature *or* a mathematical truth? Newton's answer would be that it is *both*, mathematical truth and empirical law. This answer seems strange for the modern reader, and its justification is somehow intricate even from the point of view of Newton's philosophy of science and mathematics. To sketch the main argument: Newton understood his laws of motion as most general results of induction, based on phenomena of matter in motion, *and*—backed by his theory of mathematical and absolute space and time—as mathematical truths. They are *not* hypothetical in the modern sense, i.e., in the sense of "being revisable in the light of new empirical evidence," but have to be considered as irrefutable *axioms*, as he literally writes in a letter to Roger Cotes (Newton 1959–77: V, 377).

A more comprehensive account of the set of problems involved would lead us, *inter alia*, into details of Newton's method of analysis and synthesis and what role geometry plays for him to secure empirical knowledge (see Guicciardini 2009). It is noteworthy that already Newton's designation of his first principles as *axiomata sive leges motus* (axioms or laws of motion) is revealing and indicates an important tension which became manifest later. As *leges motus*, these first principles are basic laws of nature, meant to govern the behavior of (possibly all) bodies natural philosophy has to do with—they are meant to physically *explain* phenomena. However, as *axiomata* they act as first and most general principles of the deductive structure of his theory of mechanics, i.e., they are meant to be able to organize the whole body of mechanical knowledge, including more special laws of mechanics. Philosophically it is not at all clear that one and the same set of principles can fulfil both functions. In conclusion, Newton, albeit his empiristic avowals, was not a representative of a modern, hypothetical-deductive understanding of science, but of a traditional, axiomatic-deductive understanding of rational mechanics, which can be labeled as "Mechanical Euclideanism." In this regard, Newton was, despite all his new findings, a typical representative of classical mathematical philosophy of nature before the early nineteenth century (see Pulte 2005).

The most important new finding of the *Principia* is, of course, Newton's discovery of the universal law of gravity. Its special $1/r^2$-form allowed Newton to treat (nearly) spherical bodies like planets as "mass points," i.e., as if the whole mass would be concentrated in a single point. He thus gained a set of principles—his three laws of motion and the law of gravity—to treat terrestrial motions—like Galileo's free fall or Huyghen's pendular motion—and celestial motions—like Kepler's planetary motions—from a general and unifying point of view. Already in Book 1, Section 3, he deduced for a body revolving in an ellipse that a centripetal force, directed towards a focus of the ellipse, is inverse to the square of the distance of the body and the focus (or, in more concrete terms, of planet and sun; cf. Newton [1726] 1999: 462–3). Newton, late in his life, claimed that he discovered already in the 1660s that gravity follows a $1/r^2$-law. At that time, however, he was still looking for a more or less Cartesian explanation of planetary motion, i.e., an explanation rather in terms of vortices than of a centripetal force. His later research on planetary orbits and the orbit of the moon, especially his exchange with Robert Hooke on these subjects, were decisive for Newton's most important discovery (Cohen 1999: 64–82). At the end of *Principia*'s Book 1, however, he already applied the inverse square law already to the three-body problem—one of the most demanding problems of mathematical physics until the end of the nineteenth century—and gained some qualitative results (Newton [1726] 1999: 570–88). Although in the following Scholium (ibid., 588) he described his wide use of the word "attraction," leaving open his view on its mechanical cause, from today's point of view it seems fair to claim that for Newton force "was an entity ontologically existent in the universe" (Westfall 1971: 507). One important claim of Newton then was—in his Scholium at the end of Book 2—to have demonstrated that a Cartesian explanation of celestial motions by means of vortices is impossible (Newton [1726] 1999: 786–8). In Book 3 on "The System of the World," Newton developed his *general* law of gravitation in a series of steps (cf. ibid., 802–15). In a slightly modernized form, it says that between Masses M and M' with the distance r there exists a mutual attracting force F with:

$$F \sim \frac{M \cdot M'}{r^2} \qquad (3)$$

It is *this* law and its interpretation in terms of "action at a distance" are part of the *Principia* that can be labelled "revolutionary." Newton made extensive use of it to explain the motion of the planets and the moon, the shape of the earth, the theory of tides, the orbits of comets and other empirical findings.

The *Principia* ends with a General Scholium, added in the second edition from 1713. Here we find Newton's famous, "I do not feign hypotheses" (Newton [1726] 1999: 943), which again has to be placed in the context of the (unknown) mechanical causes of gravity or, to be more specific and explicit, in his clever strategy to present his theory of gravity in an acceptable way for the proponents of traditional mechanism that rejected action at a distance as the reintroduction of "occult qualities." No less influential was his concluding outlook on phenomena such as electricity which still had to be explained and mathematized according to his principles, as well as his commitment to an understanding of science as a kind of divine service that has to take into account the divine institution of nature: "to treat of God from phenomena is certainly a part of natural philosophy" (ibid., 943).

It is mainly Book 3 on which the later fame of the *Principia* rests, and it is Newton's claim to have brought together empirical research and mathematical certainty in a new form of science that made an upcoming "Newtonianism" so attractive for many mathematicians, physicists, and philosophers in the Age of Enlightenment (see Pulte and Mandelbrote 2019 and Chapter 4). Almost a century after Newton's death, Joseph Louis Lagrange saw him not only as the greatest genius of all time, but also "the luckiest of all men, since there is only one system of the world to be discovered" (cf. Lagrange *Oeuvres*, 1867–92: vol. 1, xx). "Describing and understanding" the system of the world in Lagrange's time was not only about the world system in the sense of Book 3 of Newton's *Principia*, i.e., the planetary system, but about a unified mathematical worldview that encompasses the supralunary sphere as well as the sublunary sphere, based on the same mathematical principles. Newton paved only *one* way for this new and unified mathematical worldview. Though this way later was interpreted as a one-way main road, there were other ways developed by other scholars, some taking quite different directions.

Other Programs and Mathematical Principles of Natural Philosophy

While Newton's program soon gained ground in Britain, it faced serious resistance on the Continent. Descartes' natural philosophy did not allow for active principles or "forces" of matter but aimed at a mathematical conceptualization of mechanics on the basis of passive matter and motion only. Accordingly, the basic laws of mechanics had to be "contact laws," like the law of conservation of "quantity of motion" (or, in modern terms) of impulse $m\,v$, formulated by René Descartes in rudimentary form already in his *Principia* from 1644 and defended by his adherents in France. For the Cartesians, action at a distance—an implicit new type of mechanical explanation in Newton's program—was a philosophical impossibility and absurdity of the first order. To mention a third important program of a mathematization of nature on mechanical principles: Newton's great rival Gottfried Wilhelm Leibniz also rejected Descartes' theory of (passive) matter and tried to establish mechanics based on a complex metaphysical theory of forces (see Stammel 1982). Notwithstanding his dynamical theory of matter, however, Leibniz rejected Newton's gravitation and his theory of absolute space and time as philosophically unacceptable. His correspondence with Newton's "mouthpiece" Samuel Clarke gives an impression of the fundamental differences between the two program (see Chapter 4 of this volume). For Leibniz, active forces of matter manifest themselves only in the direct contact of two bodies. In the year before Newton's *Principia* appeared, he published a short paper, his *Brevis demonstratio*, that tried to prove against Descartes that in nature the magnitude of impulse is *not* conserved. Instead, he claimed that in all natural processes so-called *vis viva* (living force) is conserved, i.e., the sum of $m\,v^2$ (and not $m\,v$). Unlike Descartes and Newton, Leibniz denied the existence of atoms and perfectly hard bodies and believed instead that bodies are always elastic. This is the main reason for his claim to universal generality of *vis viva*-conservation. As he and his immediate followers did not develop the concept of potential, it took until the general principle of (what is today known) conservation of mechanical energy was developed (Scott 1970). The reader should keep in mind here that the idea of *conservation* of certain mechanical quantities of motion had no place in Newton's mechanics.

Although Descartes, Newton and Leibniz shared the conviction, rooted in corpuscular philosophy, that nature has to be described and understood in terms of mechanical principles, we can retrace how they arrived at very different mathematical conceptualizations of their programs. Contrary to Kuhn's view (1996: 30–3, 39–40), no "Newtonian Revolution" with respect to the foundations of mechanics happened, and for this simple reason, the first half of the eighteenth century can by no means be characterized as "normal science" in a Newtonian tradition (Pulte 1989: 10–20). It can rather be characterized as a continuing competition of different research programs. Three important controversies may illustrate this: the struggle whether *forces* at all should be allowed as basic concepts of mechanics (Cartesians vs. Newtonians and Leibnizians), the fight whether gravity can be accepted as an explanatory category of mechanics (Newtonians vs. Cartesians and Leibnizians) and the argument about what magnitude is really conserved in mechanical process, Descartes' momentum *m v* or Leibniz's living force $m\ v^2$ (Leibnizians vs. Cartesians and "later" Newtonians).

These conflicts promoted further engagement with mathematical principles of natural philosophy. Since mechanics was understood by all representatives of mathematical physics to be an essentially mathematical-deductive science, it was crucial to start from the "correct" and most general principles and try to show that rival "principles" are mere deductive consequences of the own, true principles. This is the main reason why we observe in eighteenth century rational mechanics a "mania of demonstration," as Ernst Mach ([1933] 1982: 72) put it. The safeguarding of one's own principles through philosophical and mathematical arguments—and not so much through direct empirical evidence—was of great importance in the first half of the century. We are here dealing with one "Euclidean ideal" of mechanics, spelled out in different, rival programs.

There were, of course, more reasons why the development of mechanics cannot be put in a "Newtonian straitjacket." One is that the Continent became dominant over Britain in mathematics in the early eighteenth century (cf. Chapter 3 of this volume). This holds especially for the further development of the calculus and its growing importance for mechanics. Another reason is that Newton's laws were not suitable for solving certain mechanical problems, such as the motion of rigid bodies, the motion of fluids or the oscillations of strings (Truesdell 1968). If mechanics claims to explain all nature, it cannot limit itself to problems that are reducible to point masses but also must describe and understand such phenomena. The development of continuum mechanics was mainly carried out by continental mathematicians such as the Bernoullis, Jakob Hermann, Jean le Rond d'Alembert, and Leonhard Euler.

Pars pro toto for this important field, we take a short look at Daniel Bernoulli (1700–82) and his *Hydrodynamica* from 1738, which presents the first adequate mathematical analysis of the forces and motions of an incompressible fluid. Though Daniel was strongly influenced by Newton's *Principia*, he made extensive use of the principle of *vis viva*-conservation, which he associated more with Huygens than Leibniz. His basic idea was that living force, "stored" in a certain amount of water, is never lost, but will be transferred to another amount of water, or to other material. He also made use of the premise that the total amount of mass in hydrodynamic processes is constant. Thus, he successfully brought in *conservation principles* into hydrodynamics that are not laid out in Newton's program. Daniel Bernoulli's *Hydrodynamica*, based on a few mechanical premises, would become a standard work for theoreticians and practitioners in the nineteenth century (see the section on, "The Calculus and Analytical Mechanics: The Long Road to a Unified Approach and Its Problems").

This example of the successful use of mechanical principles can perhaps illustrate what the more general development of mechanics in the first half of the eighteenth century was all about. The different programs of rational mechanics and their use to solve problems of certain classes of phenomena brought about a great number of so-called "principles" (cf. Dugas [1955] 1988), the exact status of which became more and more questionable. All these rules grew out of the study of special problems and the idealization of physical situations, they were confirmed by successful applications to different classes of phenomena, and most of them were backed by the scientific metaphysics of the respective program. However, due to their restricted scope they could no longer be understood as "principles" in a literal sense, and together did not form a coherent and, ideally, axiomatic-deductively organized corpus of mechanical knowledge. The "inflation of principles," described in more detail elsewhere (Pulte 2005), raised doubts as to whether such a corpus could be reached at all. It is mainly due to the calculus of Leibniz and Newton and its enormous expansion (calculus of variation, theory of ordinary and partial differential equations and potential theory) that ensured confidence in such an integration of the growing mechanical knowledge was not lost.

The Calculus and Analytical Mechanics: The Long Road to a Unified Approach and Its Problems

Newton, notwithstanding his co-discovery of the calculus, presented his work in geometric language. On the Continent, however, Leibniz's formulation of the calculus found early appeal among mathematicians such as Johann and Jacob Bernoulli as well as Jacob's student Jakob Hermann in Basel or Pierre Varignon and G.-F.-A. l'Hospital in France. They applied the calculus to a great number of mechanical problems. Their works were anything but a mere "translation" of Newton's mechanics into the language of the calculus, as many standard histories claim. Rather, they were original and mostly independent contributions that also followed mechanical investigations by Descartes, Leibniz, Huyghens and others. Motions of masses along curves—for example planetary motions on ellipses or thrown bodies on parabolic trajectories—were *geometrically* approximated by polygons (as Newton did) and then, by determining limiting values, treated *analytically* in order to determine tangents, deflections from tangents and thereby possibly also accelerations and acting forces. In this tradition, the (later) so-called second law of motion as differential quotient appeared for the first time:

$$F \sim \frac{d(mv)}{dt} \qquad (4)$$

Varignon, Hermann and Johann Bernoulli used it from 1700 onwards (Hankins 1967), without associating it with Newton's original second law *(2)*.

FIGURE 5.2. Engraving from Euler's *Mechanica* (1736). The image brings together the effects of gravitation in the supralunary sphere, with the planets circling around the Athena figure, and the sublunary sphere, with the pendulum on the left and 's Gravesande's *gouttière*, a pedagogical device for showing the path of a falling body, on the right. Photo courtesy of Academia Scientiarum, ECHO, Max Planck Institute for the History of Science, Library. Public Domain.

Leonhard Euler, probably the greatest champion of eighteenth century mathematics, was also a master of the use of the calculus in mechanics. Within about 30 years, he published three major textbooks on rational mechanics. The first appeared already in 1736 and was also the first to rightly claim to present mechanics in general on an analytical basis, as the title makes clear: *Mechanica sive motus Scientia analytice exposita (Mechanics, or Analytical Presentation of the Science of Motion)*. This was the first comprehensive textbook on mechanics which made extensive use of the calculus. But it was still very Euclidean in style, starting with definitions, basic propositions and proceeding by logical inferences. Euler explicitly linked his account to Newton's *Principia* and Hermann's *Phoronomia* and set out to translate their synthetical (i.e., geometrical) treatment of mechanics into an analytical one. This is one important reason why in many histories he is presented as a "Newtonian" mathematical physicists. At first glance, this seems to be apt: The basic concepts of the *Mechanica* are absolute space and time, mass and inertia (taken as *vis inertiae*, an inner force as in Newton) and power or force (*potentiae, p*), which he seems to understand as an outward and directive activity, able to change a body's motion. It is interesting to note that according to Euler *p* is determined statically, for example by the weight with which a body presses on a surface. However, its main function in the *Mechanica* is dynamical: p is the directive power that brings a body from rest to motion or changes the motion of a moving body along the line of its action. Euler presented the law of motion *(4)* as:

$$dv \sim \frac{pdt}{m} \qquad (5)$$

and claimed that this law included all known principles of motion (Euler 1736: vol. 1, 50–1). Moreover, he thought he could demonstrate that this law was "not only true, but necessary true" (ibid., 10, 56). This is another example of how rational mechanics was understood as mathematics: Mathematical truths were understood by most practitioners as self-evident and even necessary, and Euler's law *(5)* participated in this view.

The seemingly simple "piece of mechanics" *(5)* illustrated the ongoing discussion of foundational problems of mechanics: it could be interpreted in a "Newtonian" manner if Euler would interpret the force or power p as *cause* of the change of motion dv. However, this interpretation is not adopted by him at all: Euler wrote that we only know that elastic bodies and vortices can bring about forces. After he determined the magnitude the force of a (already moving) body in a bent tube on its wall, he stated that "we cannot imagine how forces can exist without motion [. . .]. We conclude that all forces we perceive in the world are brought about by motion [. . .]" (Euler 1736 I: 14–15). While Newton wanted to trace back motions to forces as its cause, Euler wanted to trace back forces to motion as its cause. He stuck to a traditional, basically Cartesian program of philosophy of nature (Pulte 1989) . This implies that mathematical representation and philosophical foundation do no longer coincide here: Euler made use of law *(5)* and of directional forces p in general mathematically, because this turned out to be empirically fruitful, but at the same time asked for a further causal explanation of forces. Throughout his long career, he did not believe in the existence of "Newtonian" forces like gravity and did not accept force as an ultimate explanatory concept.

We find here a first sign of "distancing" of mathematical representation and philosophical foundation. Another aspect of the *Mechanica* points in a similar direction: Euler was proud to present mechanics in the language of the calculus, but found it important still to illustrate his calculations and findings geometrically.

Though Euler did not use geometry any longer to demonstrate theorems, he still found it important to describe the movement of bodies in space and time in this way. The comparatively new calculus was too counterintuitive without a "bridge" to the perception of the behavior of real bodies. The same holds for d'Alembert's *Traité de Dynamique* (1743) which appeared some years later—another remarkable work, aiming at presenting the mechanical knowledge of that time based on certain and undisputable principles.

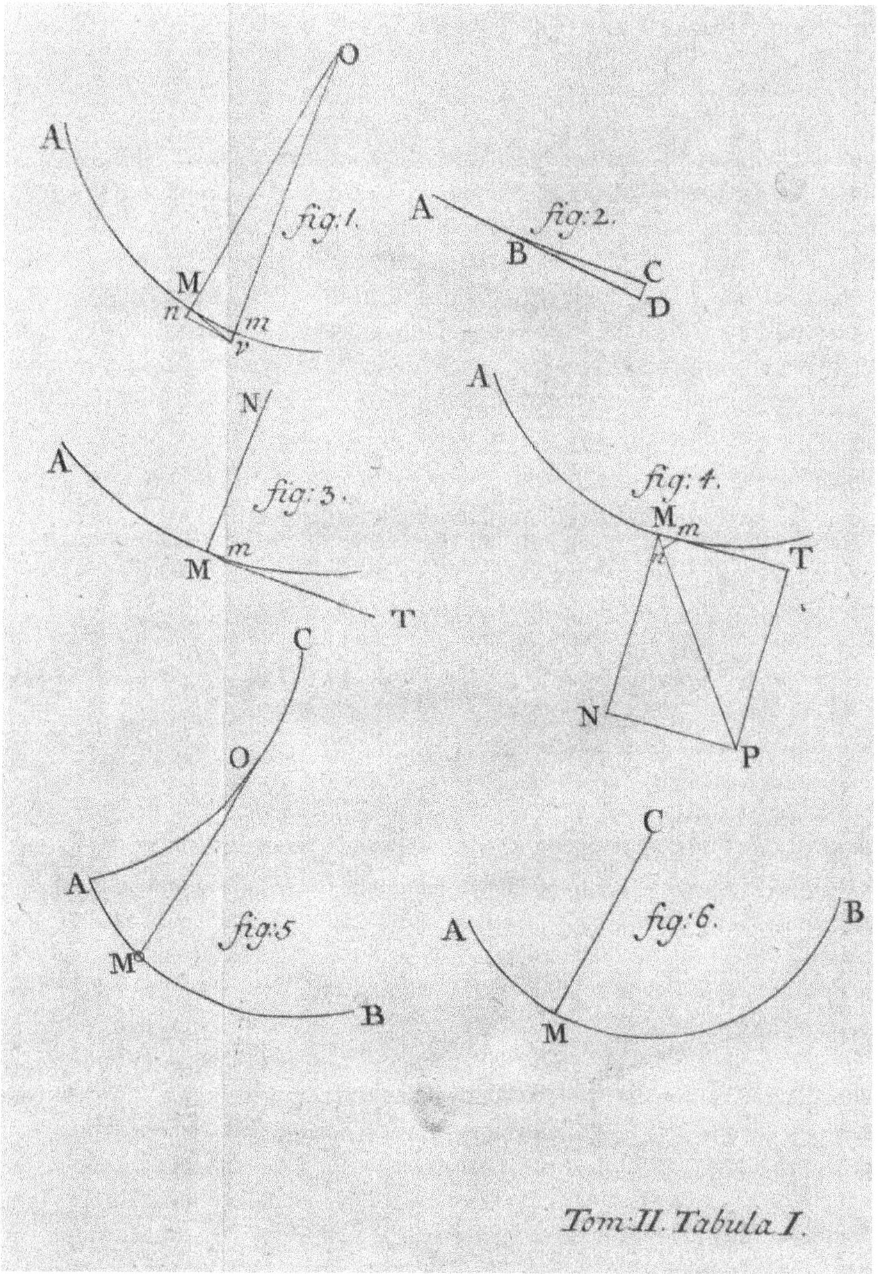

FIGURE 5.3. Geometrical illustrations in Euler's *Mechanica* (1736). Photo courtesy of ETH-Bibliothek Zürich, Rar 4689. Public Domain.

Euler was a master of applying the calculus to very different kinds of mechanical problems, making extensive use of the concept of external directive force. Using *(5)*, he formulated and solved differential equations for a variety of classes of motions in his *Mechanica* without explicit coordinate systems. The "relevant" directions are determined by the concrete problem, as Figure 5.3 illustrates. In later works, Euler introduced Cartesian coordinate-systems to treat mechanical systems—he was, indeed, decisive for the introduction of this new mode of representation. In the paper, "Découverte d'un nouveau principe de mécanique" ("Discovery of a New Principle of Mechanics," 1750), he presented for the first time the so-called second law of motion in the general form *(1)* known today. Due to the chosen units, the general equations of motion assume this form for a coordinate-system with x, y, z as axes (Euler 1750 : 90) :

$$I.\ 2M\,ddx = Pddt^2, \quad II.\ 2M\,ddy = Q\,dt^2, \quad III.\ 2M\,ddz = R\,dt^2. \qquad (6)$$

Euler was also the first to see that this law could be applied to all discrete mechanical systems (Truesdell 1968). This means that, in modern terminology, the movement of a system with *N* bodies can be represented by :

$$\underline{F}_{k} = mk\underline{\ddot{x}}, \qquad with\ k = 1, \ldots, N \qquad (7)$$

with $\underline{F}_{k}$ as the force acting on the *k*th body of the system and $\underline{\ddot{x}}$ as its acceleration. These laws turned out to be extremely fruitful when applied to more complex mechanical systems of discrete mass points, and were also used in order to treat problems of continuum mechanics by discretization.

Euler also played a crucial role in the development of "analytical mechanics." In most textbooks, this label is not used when—as in *(6)* or *(7)* above—laws are expressed by means of the calculus, rather, it is reserved for theories based on mechanical principles formulated in the language of the "higher" calculus, i.e., the calculus of variations (see Chapter 3 of this volume). Here, integral and differential principles have to be distinguished. The first ones take the form of path integrals or time integrals, corresponding to the motion of a single mass (or later, a system of mass-points) over a certain mechanical magnitude. The basic idea is that the actual value of the integral is a minimum (or, sometimes a maximum or another stationary value) and the so-called "variation" of the integral is zero. The procedure has certain analogies to the determination of extreme values of functions by "normal" calculus, for which a necessary condition is that the first derivative of the function is equal to zero. The main difference, however, is that in the case of integral principles the outcome of the variation process is not a single value (or a set of values), but the outcome can be, under certain conditions, a differential equation the solution of which is itself a function, expressing the trajectory of motion between too fixed points P_1 and P_2 in space.

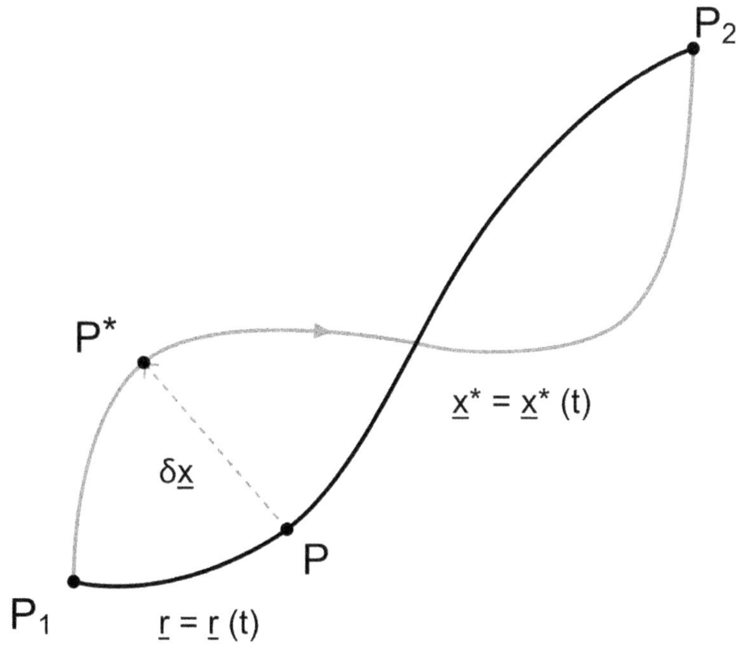

FIGURE 5.4. The Idea of Variation in the Case of Integral Principles of Analytical Mechanics. Image courtesy of Helmut Pulte.

Let $\underline{r} = \underline{r}(t)$ be the (vector) function that describes the actual trajectory of a mass under the influence of certain forces. Let other trajectories, which are thought to be possible but are not actually realized, be described by functions $\underline{x}^* = \underline{x}^*(t)$. They can be constructed by "adding" displacements $\delta\underline{x}$ to the points P of the trajectory. However, P_1 and P_2 are the same for the real trajectory and all "virtual" trajectories. If A is the mechanical magnitude thought to describe the motion, an integral principle of analytical mechanics takes this form :

$$\int_{s1}^{s2} A\,ds = Minimum\,(stationary) \quad \text{therefore} \quad \delta\int_{s1}^{s2} A\,ds = 0 \qquad (8)$$

where δ symbolizes the variation procedure that takes place here. This procedure is extremely flexible, integral principles are therefore very useful to tackle different kinds of mechanical processes. They are, however, also quite peculiar: While differential equations such as Euler's *(6)* can be interpreted as determining the path of motion successive (from "point to point"), integral principles like *(8)* seem to view, so to speak, the *whole* movement (from P_1 to P_2)—somehow *as if* the mass *m* "knew" which one of all possible paths to take to minimize the integral. Is nature wise, by being parsimonious?

Why and how did such principles enter eighteenth century's attempts to describe and understand the world mechanically? There is no simple answer, but at least one important philosophical aspect has to be taken into account here: Most philosophers of the seventeenth and eighteenth century asked for a *causal* explanation of nature on the basis of corpuscular philosophy. Thinkers as different as Francis Bacon and René Descartes therefore demanded to break with the old tradition, going back at least to Aristotle, of trying to explain nature *also* according to its so-called "purposes" or "final causes." Leibniz, however, demanded that both types of explanation of nature—causal and final explanation—be allowed and, if possible, that both be reconciled. This was *not* because Leibniz adhered to a traditional teleology of nature, still present in Newton's thinking that aimed at recognizing God in nature, but rather for "architectural" reasons. He did not believe that our cognitive capacity is sufficient to recognize, with the help of causal analysis alone, all laws of nature and to disclose their logical relationships. Teleological principles should be used complementarily to arrive at a transparent mechanical theory of nature—a theory with a clear architecture, accessible to man by means of mathematics. His sophisticated approach was reduced in a simpler and more traditional metaphysical form by Pierre Lois Moreau de Maupertuis, who in 1744 presented, with great pomp, his so-called "principe de la moindre action" ("principle of least action")—an alleged law for all natural processes, designed to show that nature is, indeed, wise and parsimonial. However, the sensation and controversies brought about by his bold claims cannot be pursued further here (see Pulte 1989).

Independently of Maupertuis, Euler had gained nearly at the same time a more general formulation of the principle of least action and published it in an appendix to his *Methodus inveniendi* (1744), the first textbook on the calculus of variation. There it actually took the form *(9)* of an integral variational principle, slightly modernized as:

$$\int mv\,ds = \int mv^2\,dt \rightarrow Min. \qquad (9)$$

Euler applied it to moving bodies under the influence of "conservative" forces. Until 1751 he also extended the application of his principle to more complex mechanical systems, including models of continuous bodies. He further considered cases where nature does not act "economically," but where the action can become a maximum or another stationary value. By introducing the concept "effort," later known as "potential," Euler made an important contribution to understanding the relationship of directive forces, conservation of *vis viva* and the general conditions under which the principle of least action can be applied to mechanical systems. Overall, Euler's *Oeuvre*, which raises both "Newtonian" and analytical mechanics to a new level and integrates the two approaches, is without parallel in the history of modern science.

The one who profited most from Euler's analytical achievements was Joseph Louis Lagrange. Aged 18, he already contacted Euler to discuss matters of the calculus of variations. He further developed Euler's approach into his so-called "δ-calculus" that was free of all geometric borrowings. He also formulated Euler's principle of least action in the language of his calculus and considered it to be "la clé universelle" ("the universal key") of all problems of mechanics, both of statics and dynamics. As early as 1760, he provided a further generalization by showing that for the motion of bodies under the influence of central forces the laws of the conservation of momentum and of the conservation of moment of momentum hold. He even applied it to problems of the mechanics of solid und fluid continua. His most important early achievement, however, was without any doubt deriving the (today so-called) "Newton's equations of motion," similar to the form *(5)*, from the principle of least action for the first time. This success showcases, as it were, the dominance of the analytical approach to mechanics over the earlier Newtonian approach, because large classes of mechanical problems could be solved without making use of the concept of an external and directive force, which several mathematicians and philosophers still found epistemologically problematical.

Lagrange's success, however, was not complete: he could only derive the equations of motion for conservative systems, i.e., by assuming the conservation of *vis viva.* But Lagrange strove for a theory of mechanics starting with as few principles as possible at the top, ideally even with only one principle. This is probably the reason why he further extended the principle of least action after 1760 and later replaced it with the principle of virtual velocities. The latter is an extension of a principle proposed by d'Alembert already in 1743 based on Lagrange's δ-calculus. It is also an analytical principle like that of least action, not an integral principle however, but a *differential* principle of rest and motion. In short, a comparison of real and only possible behavior of a mechanical system is made, but the comparison now does not extend over a finite period of time (cf. Figure 5.4), but rather the actual state of a mechanical system at a *certain moment* is compared with a neighboring state, gained by a virtual displacement $\delta \underline{r} = (\delta x, \delta y, \delta z)$ of the original position for a fixed time element dt. (Because dt is fixed, Lagrange can label the displacement "virtual velocity.") The behavior of a mechanical system is now characterized by the requirement that the so-called virtual moment δM disappears at any point in time or—in more modern and physical terms—that no virtual work is expended. For a force $\underline{F}$ acting on the system this means that:

$$\delta M = \underline{F}\,\delta\underline{r} = F_x\,\delta x + F_y\,\delta y + F_z\,\delta z = 0 \qquad (10)$$

In case of a *moving* system, we have, according to d'Alembert's original formulation, to consider not only the external directive force $\underline{F}$, but also—in modern terms—the opposite directed inertial force $m\underline{a}$ brought about by the movement, i.e., in *(10)* enters $\underline{F} - m\underline{a}$. For a system of *n* mass-points we may further extend *(10)* to the total virtual moment of all masses, being the sum of all single moments, because each single moment has to vanish. By considering both generalizations, we obtain:

$$\sum_i \left[\left(X_i - m_i \frac{d^2 x_i}{\mathrm{d}t^2} \right) \delta x_i + \left(Y_i - m_i \frac{d^2 y_i}{\mathrm{d}t^2} \right) \delta y_i + \left(Z_i - m_i \frac{d^2 z_i}{\mathrm{d}t^2} \right) \delta z_i \right] = 0 \qquad (11)$$

This is, by and large, how Lagrange presented his new fundamental principle in his masterpiece *Méchanique Analitique* (Lagrange 1788: 200, cf. 10–11). The book consists of two large parts on statics and dynamics. Lagrange obtained the dynamical principle *(11)* from a generalized form of *(10)* that holds for statics. In this respect, it can be said that he reduced dynamics to statics. On the other hand, *(11)* can be understood as including statics, simply by neglecting the "dynamical" terms. Lagrange held the opinion that he founded the whole of mechanics, both statics and dynamics, on a single analytical principle, i.e., the principle of virtual velocities. He substantiated this claim by deriving the most important laws of mechanics known at the time from this principle, including the known conservational laws and the principle of least action, which he no longer understood as a principle of metaphysical interest, but as a "simple and general result of the laws of motion."

Without any doubt, Lagrange's *Méchanique Analitique*, published a century after Newton's *Principia*, is the crowning achievement of eighteenth century rational mechanics in *mathematical* regard. The integration of the mechanical laws known at that time into a logical-deductive system is an enhancement proving the power of the calculus of variations. Since Lagrange admired Newton as the discoverer of a new "system of the world," *he himself* would have seen his masterpiece also as a culmination and perfection of Newtonian mechanics. However, this integration process refers back to a much more colorful and heterogeneous development that was anything but "normal science" in a Newtonian tradition. We therefore want to conclude this overview by looking not only at the gains of this integration, but also at its losses.

Lagrange shared the "Euclidean ideal" of Newton, Euler and others to present all mechanical knowledge in an axiomatic-deductive form. *Unification* was his overarching aim, and the laws of the various programs of rational mechanics were its objects. The price of this unification was a momentous break with the past: the *Méchanique Analitique* is the first major textbook in the history of theoretical mechanics which dismissed conceptual foundations. The systematic part of the work started directly a discussion of the analytical principles and their interrelations. Moreover, geometrical methods remained important in mechanics, but Lagrange, however, was proud to have eliminated geometry completely from his main work. He wanted to make mechanics a "new branch" of analysis (Lagrange 1788: vi). Therefore, he seemed convinced that analysis could provide a foundation for empirical science without referring to concrete experience. The *Méchanique Analitique* revealed an "analytical monism" in a threefold sense: no scientific metaphysics to ground mechanics, no geometry providing mechanics with intuitive evidence, no reference to concrete observation and experiment to gain empirical evidence. One important consequence of this remarkable *metatheoretical* shift concerns the status of his first principle. In the first edition, he characterized the principle of virtual velocities as "une espèce d'axiome de Méchanique" ("a kind of axiom for mechanics," Lagrange 1788: 12). Later, however, he dismissed this view and tried to give demonstrations of his principle on the basis of more intuitive premises—a contradiction in terms if the title "axiom" is taken seriously. Lagrange's Euclidean ambition for unification undermined traditional "Mechanical Euclideanism" as represented by Newton, Leibniz, or Euler. This led to a "crisis of principles" at the end of the eighteenth century (see Pulte 2005).

BROADENING THE PERSPECTIVE: SOME KEY EVENTS BASED ON AND BEYOND MECHANICS

The new mechanics, according to its origins neither "classical" nor "Baconian", made up the core of a mathematical worldview in the Age of Enlightenment and permeated all sciences. Describing and understanding the world went beyond this as mechanical principles were applied other areas as well, even to areas where these principles did not seem applicable, at least not at a certain point of time. We take Kuhn's distinction (cf. the beginning of this chapter) as a rough guide to some major achievements from areas already mathematized before the Scientific Revolution ("classical sciences") and from domains in the course of becoming sciences, mainly due to new experimental methods and observational findings ("Baconian sciences").

Classical Sciences

The classical sciences relevant for this chapter are astronomy, optics, harmonics and (in parts) geometry. Without any doubt, *astronomy* made the greatest progress among these subjects, because the movement of celestial bodies was directly linked to the foundational developments of mechanics. The focus was on calculating the movements of the planets and the moon, including mutual disturbances. Mathematical challenges also arose in physical astronomy, first and foremost, the problem of the shape of the earth.

Newton, in his *Principia*, had argued that the earth, due to its rotation around its own axis, was not approximately spherical, but flattened at the poles and broadened at the equator (oblate figure). His view was supported indirectly by pendulum experiments, direct, geodetic verification did not take place until after his death. Descartes had claimed based on his vortex theory of planetary motion, the exact opposite, i.e., that the earth must be flattened at the equator and broadened at the poles (prolate figure). The question was considered by many as a kind of "observatio crucis" between the Cartesian and the Newtonian system. With the support of the French government, an expedition under the leadership of Louis Godin was sent to South America to measure the earth's figure at the equator (1735). A second one, headed by Maupertuis and Alexis Clairaut, was sent to Lapland to execute meridian measures there. The latter expedition came back earlier with accurate measurements vindicating Newton's view, the former expedition confirmed the results. This outcome was a first major success for Newton's theory of gravitation on the continent—and due to Voltaire, Maupertuis became famous for his success as the "flattener of the earth."

Clairaut pursued the study of the shape of the earth in his *Théorie de la figure de la terre* (Theory of the Shape of the Earth, 1743), treating it as a homogeneous incompressible fluid and introduced "exact differentials" in order to determine conditions of equilibrium. Later investigations by Lagrange, Adrien-Marie Legendre, and Pierre-Simon de Laplace tied in with Clairaut's results and provided important contributions for the further development of potential theory (Todhunter [1873] 1961).

Clairaut also played an important role in another challenge facing Newton's theory of gravitation. The study of the moon's motion was an integral part of Book 3 of the *Principia*, and one of the most difficult parts, because both the gravitation of the earth and the sun had to be taken into account. It seemed, however, that the period of the apogee, i.e., the time it takes to return to the same point, could not be correctly approximated by Newton's $1/r^2$-law of gravitation *(3)*. This made Clairaut, Euler and others question whether Newton's law was actually correct in the 1740s. Clairaut therefore proposed to add an additional term, but later changed his mind. He found that one could stick to Newton's simple form if higher order contributions of the solar perturbation were taken into account. Other inequalities, for example those observed in the motion of Jupiter and Saturn due to their mutual interaction, could also be explained by similar, though more refined methods. Finally, Clairaut, supported by Joseph Lalande successfully predicted the return of Halley's Comet around the turn of the year 1758 to 1759.

At the end of the century, two great works by Pierre Simon Laplace appeared, *Exposition du système du monde* (The System of the World, 1796) and *Traité de mécanique céleste* (Treatise on Celestial Mechanics, 1799–1805, with later supplements), representing a climax and conclusion in astronomy. The first work was more qualitative and dealt with the laws of motion, celestial motions and the history of astronomy for a wide readership. The second work, on the other hand, presented astronomy at the mathematical height of the time, including topics of physical astronomy such as the figures of planets, the earth's atmosphere, or the tides of the seas for different regions of the earth. Laplace showed himself in this work to be a master in calculating the mutual disturbance of planets. Following mainly Euler and Lagrange, he made use of infinite trigonometric series to successively approximate their real orbits. He approximated the mutual perturbance of Jupiter and Saturn successfully by making use of the powers of the eccentricity and the inclination of these planets (Wilson 1985). The mathematical analysis of special perturbations led Laplace to the question whether all disturbances in the planetary system could be periodic and therefore whether the entire system could be proven to be stable in time. Building on work by Lagrange from 1778, Laplace arrived at a positive answer in the first volume (Book 2) of the *Traité*. Although his "proof" was not without flaws and was repeatedly contested until Henri Poincaré arrived at a satisfactory answer a century later, the stability of the planetary system was an integral part of his astronomy and his general worldview. It can be linked to the story about the conversation with Napoleon Bonaparte. Allegedly, Napoleon asked him where in his world system God appears. Laplace's answer is reported as follows: "Sir, I did not need this hypothesis." While Newton, a century earlier, saw the world system depending on God, and backed his science by physico-teleological arguments, the further mathematization of this system turned the tables. This is not without irony, because the physics developed by Laplace and his student (and later successor) Simon Denis Poisson was the hallmark of Newtonianism at the turn of the century.

In optics, partly mathematized already in antiquity, there are two fundamentally different approaches in the eighteenth century. Christian Huygens presented in his *Traité de lumiere* (Treatise on Light, 1695) a theory in which light was basically understood as a phenomenon brought about by small and elastic parts of an all-pervading aether. If by some cause the aether is set into vibration, these parts produce elementary light waves that propagate in aether according to "Huygens' principle," as it is called today. This wave theory was further developed by Euler and others. Newton, however, presented a different story about light in his second major work, i.e., *Opticks: Or, a Treatise on Reflexions, Refractions, Inflexions and Colours of Light* (1704), which summed up his investigations from 1672 onwards. Quite different from the *Principia*, this work was more experimental than mathematical, and more "hypothetical" (see, e.g., the final "Queries") than categorical. His work became a powerful voice in the dominant chorus of those who advocated a corpuscular theory of light. According to Newton, light was composed of small particles of different types belonging to a certain color of which white light is composed of. The mathematical treatment of light remained limited to the application of elementary geometrical methods using two quantitative concepts, i.e., refrangibility and reflexibility. Though it met with resistance on the continent, Newton's *Opticks* was widely accepted in the second half of the century (Shapiro 2019).

Newton's *Opticks*, however, did not get a status comparable to the *Principia* as a paradigm of scientific research. Once again, Euler did not follow in Newton's footsteps. His *Nova theoria lucis et colorum* (A New Theory of Light and Colours, 1746) contained a critical investigation of Newton's approach and further developed the wave theory of light. In this work monochromatic light was probably for the first time characterized as a wave with a determined frequency. Euler's three-volume *Dioptrica* (1769–71), provided probably the most advanced mathematical treatment of optics in the entire century.

Johann Heinrich Lambert, a mathematician of great philosophical weight, linked mathematical to physiological optics in his *Photometria* (1760). Following Pierre Bouguer's *Essai d'Optique* (Essay on Optics, 1729), his main interest is the luminosity of light and the measurement of its intensity. By differentiating between the absolute luminosity of luminous objects, the luminosity of an illuminated object and the subjective luminosity perceived by the eye, he obtained several interesting mathematical results. To mention only one: the illuminance of an object, located at a distance r from a point source of light, is proportional to the inverse square $(1/r^2)$ of the distance.

Harmonics, as the last among the classical sciences, or *theory of music*, to use a more modern label, was certainly not at the center of the eighteenth century's efforts at mathematization, but it should not be passed over entirely in silence either. Since the Scientific Revolution, it is strongly shaped by experiments and observations with instruments, and in so far more a Baconian science. Generally speaking, the tendency was to explain the Pythagorean integer ratios of consonances by invoking physical properties and by model-like treatments of the motion of the components of solid bodies (such as a string) and of air as a pendulum connected to each other. This is related to two areas of mechanics: acoustics, which has been pursued based on corpuscular philosophy, and the continuum mechanics of solid bodies, in particular oscillation problems of linear continua, such as the famous vibrating string problem (cf. Chapter 3 of this volume). The mathematical analysis of this problem by developing "harmonic functions" and musical harmony are closely connected (Darrigol 2007). While for d'Alembert the solution of the string problem was mainly a mathematical challenge, i.e., solving the wave equation as a quite new type of partial differential equation, Euler and Daniel Bernoulli also worked out the physical implications of its possible solutions. Bernoulli concluded that all constituents of a sound result from its harmonics on the fundamental, which allowed, for example, an explanation of the overtones of the vibrating string. Earlier Euler developed a theory of music that paid tribute to the "traditional" ratios of the consonant intervals. Though he gave a physical interpretation of these ratios on the basis of vibrational frequencies, the numbers itself formed the structure of this theory, first presented in his *Tentamen nova theoriae musicae* (New Essay on Musical Theory, 1739). The *Tentamen* did not remain Euler's only foray into music theory, but his contributions were little noticed. Arguably, his attempt to explain music and even its euphony by mathematics was too mathematical for musicians and too musical for mathematicians (Pesic 2013).

Baconian Sciences

In the Age of Enlightenment experimental sciences try to find general mathematical regularities and laws of nature through experiment, observation, and generalization by induction. The importance of the quantification of properties and the discovery of mathematical relations among them was an ongoing point of discussion (Hankins 1997). Mechanical explanations were sought after, but often out of reach at the time, thus the obtained mathematical uniformities remained on a phenomenal level. Theoretical reduction to mechanical principles was left to the future. In areas, where "certain" laws seemed unattainable, for instance in the understanding of the social world, or when dealing with unavoidable errors of measurement and observation in astronomy or geodesy, statistical und probabilistic methods became important (cf. Chapter 4 of this volume). A few remarks on even fewer, exemplary areas must stand here *pars pro toto* for these efforts of mathematization.

Gas and Heat The seventeenth-century gas law of Boyle-Mariotte, stated the inverse proportionality of pressure and volume of a given amount of gas for isothermic processes From hypothetical premises Newton, in Book II of his *Principia*, deduced mathematically that for an elastic fluid composed of particles at rest and repelling each other by a force anti-proportional to their distance, density and pressure are proportional, what amounts to the same law if air is treated as an elastic fluid. Daniel Bernoulli gave a more appropriate mathematical treatment in his *Hydrodynamica* by treating gas consisting of small, spherical and elastic particles which move straightforward with a high velocity v. He showed that the particles, enclosed in a vessel, exert a pressure p on the walls proportional to v^2, a central statement of the main equation of the later kinetic gas theory. He also *postulated* that this magnitude is proportional to the temperature T, thus opening the possibility of measuring its magnitude by pressure. The relation $T \sim v^2$ also implied that heat and mechanical living force are closely related—an idea not fully developed until the later nineteenth century. He linked his considerations about temperature and pressure to the earth's atmosphere, including empirical approximations how temperature and height above the earth are related. It was only in 1801–2, that John Dalton, Joseph Gay-Lussac and Jacques Charles would make substantial progress in the empirical determination of gas laws.

Determining these laws went hand in hand with the development of appropriate measurement devices. The thermometer for the measurement of temperature, developed by Daniel G. Fahrenheit, Anders Celsius, and others, was quite precise, but the quantification of heat encountered other fundamental conceptual and practical problems. In the tradition of Bacon and Descartes, a minority of the natural philosophers considered heat to be nothing but the motion of the smallest particles of matter, such as molecules. The majority, however, considered it to be a substance in its own right, such as a very subtle fluid that can also penetrate bodies. This disagreement was still discussed by Laplace and Antoine Laurent Lavoisier in their famous *Mémoire sur la chaleur* (Essay on Heat, 1783), without taking sides. Important for the development of calorimetry was the development of first formulas for the mixture of unequally heated quantities of a certain liquid. Georg W. Richmann gave in 1757 this formula for the mixing temperature for two quantities of liquid with different masses and temperatures:

$$T = \frac{mt + m't'}{m + m'} \tag{12}$$

Jean A. Deluc and later Johann C. Wilcke and Joseph Black saw that such a simple formula did not apply when you mix water and ice. Rather, the mixing temperature (at equal masses) was much closer to the temperature of the ice. The explanation was that heat must be stored in the ice and set free in the process of melting without considerable increase of temperature—"latent heat," as Wilcke and Black called it. Helped by the ice calorimeter of Lavoisier and Laplace, this led to a more precise determination of heat quantities—depending on mass, temperature, and specific matter constants—at the end of the century. Though measurements by Benjamin Thompson (Count Rumford) made it unlikely that heat was a ponderable substance, the fluid theory of heat remained influential. Mathematically more sophisticated investigations on heat diffusion by Jean-Baptiste Biot and Jean B. J. Fourier already belong to the nineteenth century. Testifying the unclear nature of heat, Fourier's mathematical treatment remained purely phenomenal and was later declared a paradigm of mathematical physics in general by positivists such as Auguste Comte.

Electricity and Magnetism Natural philosophers were confronted with the problem of action at a distance, similar to gravitation, though not "universal" and mass-dependent. Newton had opened the way for such an explanation by claiming that next to gravitation other attractive or repulsive forces may interact between atoms. However, early in the century, most scientists on the Continent, and even many "Newtonians" sought refuge in an "explanation" which Newton himself had left open for gravitation in some hints of the *Principia*: aether. It was understood as a subtle and elastic fluid that caused attraction or repulsion of magnets or electrified bodies, either by motion or by certain other "physical" properties, like gradients of pressure, for example. Other approaches made subtle fluids themselves the bearer of magnetism or electricity ; in the case of magnetism these were generally understood as circulating streams of magnetic fluid (see Cantor and Hodge 1981). Although from the modern point of view these fluids seem to be suspicious and superfluous constructs, their importance for the development of quantitative concepts and mathematical methods should not be underestimated. They became superfluous only when mathematical laws of physics were reached. In contrast to gravity, no "direct" measurement of electric and magnetic forces was possible at first. Newton guessed on the basis of rough observations that the magnetic force decreased "not at the square but almost as the cube of the distance" (Newton [1726] 1999: 810). More accurate measurements of magnetic forces were attempted in the first two decades of the eighteenth century, but improved experimental techniques made electricity the preferred field for the search of regularities. The Leyden jar became an important tool for charging bodies to different degrees and studying their effects on other bodies. Electroscopes (resp. electrometers) were developed in order to quantify the "charge" of electricity. Following Benjamin Franklin, Franz Aepenius mathematized electric forces in his *Tentamen* on electricity and magnetism from 1759. He presented both electric and magnetic interactions as action at a distance-phenomena, leaving open its ultimate cause. For electricity, he postulated different kinds of forces according to the phenomena of attraction and repulsion and the material corpuscules involved (Home 2019). He also stated that the forces of electrical interactions were proportional to the electric fluid bodies were charged with. He speculated by analogy about a $1/r^2$-law, though generally he left this problem open for further research. This became possible with more accurate measuring techniques, especially with the invention of the torsion balance by John Michell, Henry Cavendish and Charles A. de Coulomb. The torsion balance can be seen as a prime example of the development of precision instruments, which were increasingly developed from the later eighteenth century onwards to uncover mathematical laws of nature.

By the arrangement shown, Coulomb carried out his precision measurements of electrostatic forces between 1785 and 1791, resulting in the formulation of the law named after him:

$$F \sim \frac{Q \cdot Q'}{r^2} \tag{13}$$

Q and Q' are the respective electrostatic charges, and *r* is the distance of the centers of the spheres that are charged. Force *F* is repulsive, when both spheres hold the same "electrical fluid" and attractive, when they hold opposite

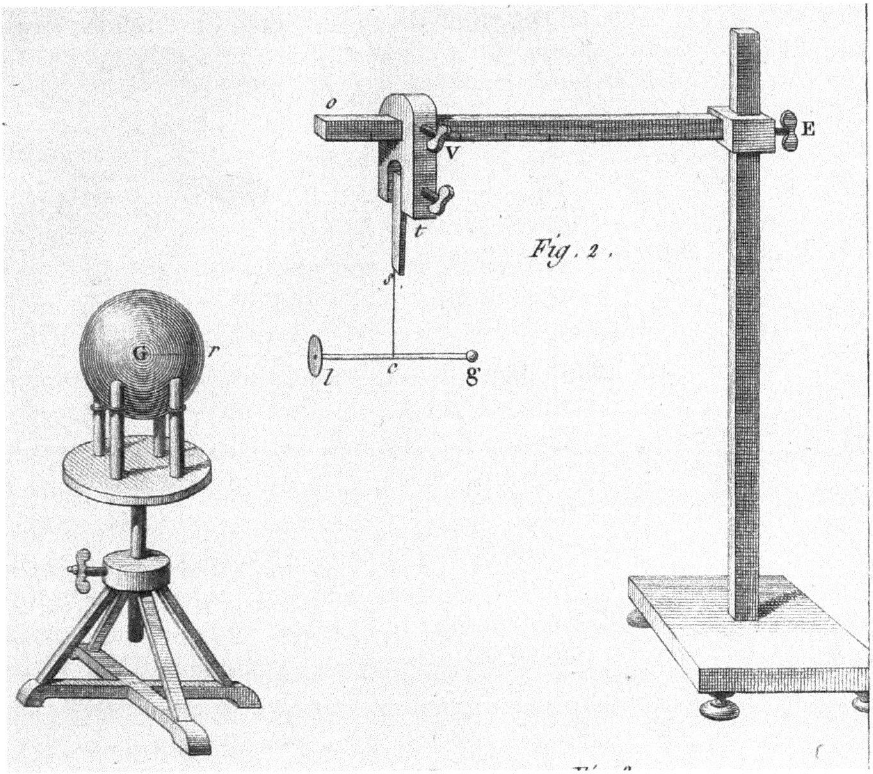

FIGURE 5.5. Coulomb's experimental arrangement from 1788 for measuring electrostatic forces. Photo courtesy of Smithsonian Libraries / Internet Archive.

"electrical fluids." The analogy to Newton's law of gravitation *(3)* obviously guided Coulomb.

Capillarity This *seems* to be a quite incidental area of phenomena, and its satisfactory mathematical treatment already leads over to the nineteenth century. However, it was important theoretically as a model of mathematical physics assuming tiny particles with microscopic forces acting between them—an extension of the Newtonian program to the realm of molecular physics. In fact, Newton already addressed the rise of fluids in capillary tubes in the last Query of the second English Edition of his *Opticks*. One main obstacle to treat the rise of fluids mathematically was estimating the range of the attractive forces of cohesion responsible for the measurable capillary effects like the height of rise or the form of the meniscus surface. Measurable, macroscopic effects on the one hand, assumed microscopic causes, *nearly* infinitesimal molecules and extremely short-ranging forces on the other hand. This challenge attracted many mathematical physicists at the turn of the century and beyond

(Rüger 1985). The most prominent among them was Laplace, who dealt with it in a supplement to the fourth volume of his famous *Traité de mécanique celeste* (Treatise on Celestial Mechanics) from 1805. As all physical phenomena at the end might be reduced to forces acting at a distance, according to Laplace, he attacked the difficult problem of capillarity to demonstrate its capacity. He assumed, without specifying the function f(r), extremely short-range forces between the molecules of the form:

$$F(r) = f(r)\, e^{-tr} \qquad (14)$$

with a very large parameter *t*. The advantage of this form is that, although the molecules are taken at discrete mass elements, acting at a distance on neighboring molecules, all effects become independent of the exact distribution of the elements, which facilitates integration processes. According to *(14)*, the discrete molecules "mimic," so to speak, a homogeneous continuum of matter. By integration Laplace determined the whole force acting on an element of the capillary surface, and by certain stability considerations he derived its differential equation and mathematical solutions for the capillary height and other empirical results. Laplace's achievement was later praised by Carl Friedrich Gauß as "one of the finest works of this great geometer."

CONCLUSION: A SHORT RETROSPECT AND OUTLOOK

Though mathematical physics on the basis of mechanics dominated in the later eighteenth century, this chapter showed that describing and understanding the world on a mathematical basis was much richer, more heterogeneous and pretty colorful. Mechanics was at the core of this project and was its most successful part. The change from a geometrical, synthetic style to analytical mechanics had to do with the emergence of new mathematical methods, often intimately connected with mechanical problems. However, it also had to do with the need to mathematically unite mechanical laws, stemming from quite different philosophical programs with quite different conceptualizations of mechanics. Their unifying bond with respect to natural philosophy was a common background in early modern corpuscular philosophy. Their unifying background with respect to philosophy of science was "Mechanical Euclideanism," i.e., the striving for an axiomatic-deductive organization of all accepted mechanical laws. Here, different and hardly reconcilable interests meet: laws have to explain nature, axioms have to organize theories. There was a price to pay for this conflict, as became clear in the case of the late Lagrange: loss of "mathematical" evidence and certainty, which characterized the earlier development. The problem was identified and pursued by later mathematicians and philosophers and was an ongoing point of a foundational discussion in the nineteenth century.

The enormous successes of rational mechanics, especially celestial mechanics, were often, but wrongly, attributed in the nineteenth century to Newton's program alone. These successes led at the end of the eighteenth century to the ideological aberration of extending "*the* mathematical method" in a somehow naïve manner to completely different areas such as social phenomena. They also gave rise to profound philosophical inquiries into the questions of why mathematics was so successful in describing and understanding the world, i.e., why mathematics contributed so much to the clarification and stabilization of our knowledge of the world. Kant, in this time, still used "Science" in the singular form, because he believed in *one and only* science based on rational mechanics. At the core of his famous assertion that "in any special doctrine of nature there can be only as much *proper* science as there is *mathematics* to be found therein" is his conviction that without mathematics no certain knowledge of nature can be gained at all—and that certainty of knowledge is what we are striving for. But as we have seen in examples of the mathematization in the "Baconian sciences"—which was part of the program from the early Enlightenment onwards—they could be realized only partially by mechanics. This remained, to use another phrase from Kant, a "focus imaginarius" of reason. The upcoming of more phenomenological approaches to mathematical physics, like Fourier's theory of heat, have to be seen in this context. Later developments, however, show a reinforcement of mechanistic thinking in thermodynamics, electrodynamics, and other areas of physics and science in general. This can be understood as a heritage of the Enlightenment: the "rational dream" of a comprehensive and transparent understanding of the world by means of mathematics.

CHAPTER SIX

Mathematics and Technological Change

Sublime Mathematics and Practical Application

JANE WESS

Developing industrialization and imperialism changed the nature of the physical and cultural landscape in Europe during the eighteenth century. For both industrial and imperial purposes mathematics was increasingly applied to "technology." Technology is not a contemporary word, but can be understood as covering aspects of life which involved the use of tools and machines (also see the Introduction in Volume 1). The applications of mathematics in the century following Isaac Newton's *Principia* can be divided into those which involved large numbers of people becoming mathematically competent in relatively straightforward processes, and those which involved the introduction of the new mathematics of the seventeenth century, in particular the calculus, or as Leonard Euler termed it "sublime mathematics." Focusing on Western Europe, and in particular Britain, this chapter touches on a range of activities which were mathematized during the period.

The formation of the European nation states in this period required exploitation of the environment, which incorporated quantification and rationalization. Europe's population was growing, a market economy had developed, and the increasing pressures on resources gave rise to new attitudes towards their efficient use. The needs of an industrialized society were considerable: surveying for new land use, improved agriculture to feed an increasing population, civil engineering, efficient transport systems, and water supply on an unprecedented scale. Power sources of men, animals, wind, water, and later steam engines, needed to be evaluated and compared. The requirements of imperialism were equally diverse. Navigational techniques were extended from the local surveys of coastlines based on the largely Portuguese "portolans" (see Chapter 5 in Volume 3) to meet the demands of those sailing further afield. Expanding rival nation states were in competition, hence the need for effective artillery, which was increasingly mathematized to optimize results. To pay for this expansion and increasing expenditure on wars, taxes on alcohol had been introduced in England and France in the seventeenth century. Assessing this tax spawned an army of excise men performing quite complicated mathematics by the eighteenth century.

Mathematical literacy was thus becoming a requirement for significant sections of society as the demand in terms of mathematical skills grew. The fact that mathematics was beginning to play a part in a wide variety of occupations is voiced in Richard Neve's *Mathematicks Made Plain* (1708), a text aimed at: "gentlemen, military officers, engineers, architects, seamen, surveyors, usurers, shopkeepers, and artificers such as masons, bricklayers, carpenters, shipwrights, watch makers, glaziers, painters, and 'plaisterers'" (ibid., title page). Mathematics was assuming its omnipresent role familiar today, permeating many aspects of everyday life.

Alongside the above development, relatively new branches of mathematics were starting to be applied to a range of physical problems, though, as will be evident from this chapter, there were often few tangible benefits. A number of leading mathematicians and mathematical practitioners were keen to apply the new, cutting-edge calculus, which had been proposed in different forms by Newton and Leibniz, to many of the physical problems that presented themselves. Euler, the most prolific mathematician of the eighteenth century, considered anything short of the new calculus to be inadequate for the optimal understanding of all machines. "The common branches of mathematics cannot lead to any very sublime knowledge. We are beholden to sublime mathematics for the modern discoveries which have been made." Euler cited artillery, navigation, hydraulics, and mechanics as requiring this "sublime" form of mathematics (Euler 1798: E720).[1]

However, in practice, the application of the calculus to technology was not straightforward and its results often contradicted by experiments and experience. D'Alembert writing on fluid mechanics in the mid eighteenth century was all too aware of the problem:

> If the case is such that the issue to be examined is too complicated for all its elements to be introduced in the analytical modeling desired of them, the most inconvenient ones are left aside, replaced by others less bothersome, but also less real, and then one is astonished at arriving at a result that nature disproves. (d'Alembert 1752: x–xj)

This led both to many heated scientific argument and to frictions between practitioners in the field and theoreticians in the academies and schools.

THE REQUIREMENTS OF EARLY INDUSTRIALIZATION

Land Management and Construction

The most fundamental change required for industrialization was that of land use. The agricultural revolution in eighteenth century Europe has been a matter of contention, but it is indisputable that between 1650 and 1750 there was a significant increase in all types of food produced, paving the way for the enormous social changes which took place at the end of the period and later. Underpinning agricultural production was land "improvement," usually involving enclosure, which reached a peak in the 1760s in Britain and continued into the nineteenth century.

Surveying was fundamental to enclosure both in Europe and North America, spawning a profession of land measurers (see also Chapter 2). This new market was provided with a wealth of textual advice, and it is noteworthy how often the books advocated a particular physical instrument. Thomas Abel of Philadelphia suggested a slide rule for calculation in 1761, Benjamin Martin in 1766 wrote on the "goniometer," Thomas Whittell and John Duncombe introduced the "dendrometer" in 1768 for finding the heights of trees. In 1779, Benjamin Talbot described a "catoptric sextant," which was simply a sextant, all sextants being "catoptric" in that they involve reflection. George Wall, again in Philadelphia, advocated a "trigonometer" in 1788. It has been suggested that this emphasis on instruments was in order to provide proof or reassurance that mathematics dealt with solid and substantial matters; that it was not merely theoretical speculation but geared towards action (Gerbino and Johnston 2009: 114).

Agriculture was not overlooked by those wishing to mathematize the understanding of processes. Arthur Young wrote in *The Farmer's Letters to the People of England* in 1771 that the present age was too enlightened to hold agriculture in contempt. Even Euler wrote on agricultural topics (Euler 1767: E341). Jethro Tull, known for his invention of the horse-drawn seed drill, lamented in *The New Horse-Houghing Husbandry* in 1731 that little had been written on the plough. He complained that makers of ploughs were unable to write their rules mathematically or show how the mechanical powers could be applied.

Half a century later James Small addressed Tull's concerns by publishing mathematical formulations to advocate the most efficient shape for the moldboard of the plough in 1784. Small suggested mathematical methods in order to produce the optimum shape for a consistent turnover of soil. Small's attitude was typical of Enlightenment thinking. He complained that, because the moldboard had been considered simple, it had been left to country artists

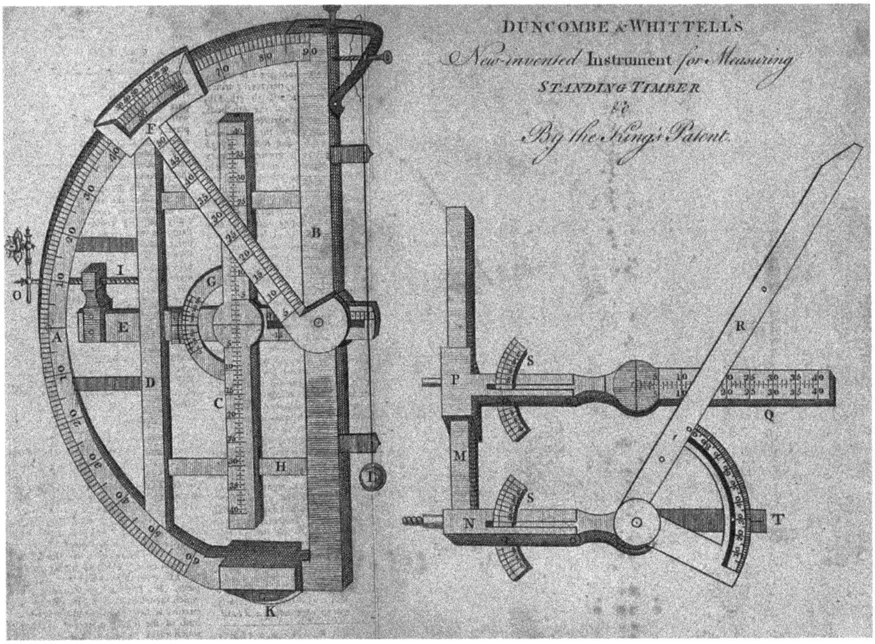

FIGURE 6.1. Duncombe & Whittel's dendrometer for measuring standing timber. Patent 1768. Credit: A dendrometer. Engraving by G. Daws. Wellcome Collection. Public Domain Mark.

who had no mathematical skill to guide them, whereas he had produced moldboards with absolute certainty, according to mathematical principles. Mathematics as embodying incontrovertible principles was a constant theme of those seeking to establish credibility by means of their mathematical knowledge. Small's moldboard became popular at the end of the century.

The nobility and the gentry were urged to act like merchants by raising capital and improving their land. For example, John Mills wrote in *A New and Complete System of Practical Husbandry* (vol. 2) in 1763 that the Marquis of Turbilly had doubled the number of inhabitants on his estate and quadrupled its produce, which he claimed was a truly great acquisition for France. Clearing woods was a necessary starting point especially in Germany and Switzerland, often using machines based on the simple powers, such as the lever mechanism invented in 1759 by the Swiss Peter Sommers. This machine was demonstrated near Berne, where it was shown to be able to lift the roots of a large oak from frozen ground in eight minutes. Drainage was an important activity undertaken from the seventeenth century across Europe, particularly in Italy and the Low Countries, but also in France, Germany, and Denmark.

Increasing populations and growing cities required a great deal of construction. Those in the building trade, like their counterparts in surveying, were encouraged to use instruments in the mathematization of their work. Henry Coggershall targeted carpenters and glaziers in *The Art of Practical Measurement by a Two Foot Rule.* The first edition of this book (1677) introduced an early version of the logarithmic slide rule. While Coggershall's book dealt with a number of more advanced topics—solids of different types, tapering shapes, casks, gauge points, ship's capacities, dialing using sines, tangents cosines and logarithms—the simpler parts of the work were probably the most in demand. One side of the slide rule could be used for multiplication of sides to find areas, the other to find volumes using the length of a side if the wood had a square cross-section, or if the wood had a circular cross-section a "girt" or a quarter of the circumference.

The significance of these measurements had been considered by the French engineer and mathematician Antoine Parent in 1708 when he suggested that the load carrying capacity of a beam was in direct proportion to its width and the square of its depth. However, these considerations did not appear in English texts until the *English Dictionary for Builders* of 1734. Peter Nicholson, who wrote books for carpenters and joiners at the end of the period, provided beautiful geometrical plates on the construction of groins, arches, and staircases, but his calculations for the strength of beams, based on those by earlier writers, were deemed inadequate (Yeomans 1999). As there is no evidence that Nicholson was responding to any need by builders of a more scientific approach, the mathematization had little consequence even by the end of the eighteenth century.[2]

On a higher plane mathematically and socially were the calculations undertaken by Christopher Wren for the dome of the new St. Paul's Cathedral after the fire of London. This has been described as one of the earliest examples of mathematics being introduced before the design process, and being applied in the actual construction of a building. Wren's design was initially based on an idea of Robert Hooke's. The catenary, or curve formed by a hanging chain, was derived mathematically in 1691 by Johann Bernoulli, Christian Huygens, and Gottfried Wilhelm Leibniz (see Chapter 3). Hooke had suggested that this shape inverted would be perfect for an arch, in particular for the middle of the three layers of the dome. However, the construction of St. Paul's in 1704 actually used a different mathematical curve, it was part of a cubic that resembled a parabola. The construction was based on the dome of Les Invalides in Paris, which was designed by Jules Hardouin Mansard and completed shortly before St. Paul's (Gerbino and Johnston 2009: 98).

Improvements to private estates entailed construction of grand houses on a scale previously unknown. An author who was increasingly using mathematics to help wealthy people improve their houses and gardens was Batty Langley, who in 1726 mathematized every aspect of good taste. Very detailed instructions concerning the proportion of columns of various architectural orders were accompanied by suggestions for geometrical walks and designs for sundials

FIGURE 6.2. Study-design for a dome with the profile of a "cubic parabola," drawn by Wren, *c.* 1690. © The Trustees of the British Museum, 1881-06-11-203.

placed on the surfaces of mathematical solids. Architects began to use scale drawings as planning tools and proportional design methods as standard practice. The Palladian style, originally developed in Venice by Antonio Palladio in the sixteenth century, became very popular in Britain, with geometry seen as providing rationality to many aspects of construction (Arciszewska 2010).

There was a particular fascination with the five classical orders as defined by Vitruvius, rediscovered in the sixteenth century. The five orders: Ionic, Doric, Corinthian, Tuscan and Composite, were distinguished by the proportion of their columns, which were laid down by Palladio. Instruments usually based on the sector, which is a hinged rule giving proportional lengths, were devised from the seventeenth century to make the reproduction of these proportions easier, although the resulting instruments were fairly complex. The Ionic order uses volutes, or spirals, which George III of England studied when he was Prince of Wales; beautiful volute compasses can be found in his collection of instruments (Morton and Wess 1993).

Arches generally, and bridges in particular attracted mathematical attention, although this had less to do with aesthetics than with utility. The fundamental issue for masonry bridges during the eighteenth century was that of the thrust against the abutments, which in turn led to other considerations of safety (Heyman 1982). Of interest were the minimum requirements for the arches, and the necessary thickness of the abutments and piers. Phillippe de la Hire, a notable mathematician and engineer, wrote a mathematical treatise on bridge design in 1695 introducing polygons of force. In 1712, he extended his study to include friction. By describing the forces at a few sites and using the concept of hinges of collapse, he was able to write equilibrium equations for the forces.

Going beyond the empirical tables found in most texts of the time, Pierre Couplet, who was involved with the Académie Royale des Sciences through his father, wrote a series of treatises between 1726 and 1730 on supporting walls and arches. Like de la Hire he moved from the frictionless case to the inclusion of friction, describing the former case as "merely mathematical and in reality false" (Heyman 1976: 225). He also defined the safety of an arch: a semicircular masonry arch will not collapse if the chord drawn from its top center to the bottom of the outside of the arch remains inside the masonry structure. In France the importance of bridge construction was recognized with the establishment of the Écoles des ponts et chaussées in 1747. By the end of the period, Jean-Rodolphe Perronet, who taught at the Ecole, was able to use the opposition of forces of bridge arches of similar spans to reduce the size of the abutments and create bridges which are now considered the zenith of masonry arches.

In some instances, builders were not able to follow mathematical theory. Henri Gautier, who had written on the practicalities of construction of bridges in 1716, had, the previous year, asked de la Hire for mathematical directions, but found them useless as they contained algebra (Ruddock 2017: 64). The person revered as the first civil engineer in Britain, John Smeaton, considered optimum forms for sustaining forces. For Blackfriar's Bridge he suggested circular segments of 120 degrees: "an arch which has the tendency to persevere in its original figure is stronger than a much thicker arch" (Ruddock 1997: 157). In general, Smeaton's bridges used no sophisticated mathematics, nothing "sublime." It would be satisfying to record that Smeaton had established the shape of Eddystone lighthouse, which became the Platonic shape for a lighthouse, using mathematics, but according to him he did not. Only later experiments in the nineteenth century demonstrated mathematically the success of Smeaton's empirical design against the ravages of the waves.

Trade and Transport

Various trades depended on accurate weighing. For example Charles Sommers wrote *A Treatise on the Method of Weighing Gold and Silver* (1756) on the use of hydrostatics, when volume and weight could be compared by submersion in order to prevent counterfeiting. He cites the work of Robert Boyle and John Theosophilus Desaguliers, both well-known experimental philosophers, advertising his hydrostatic balance apparatus. He also provided balances for Spanish and Portuguese gold coin, for currency from the West Indies, scales and weights for Turkey, the East Indies, Venice and Guinea. Hydrostatic balances, involving weighing items in liquids, are useful for comparing the density of materials. The mathematics involved in these calculations was long familiar; however, the new calculus was introduced into other applications of weighing.

The textile trade, and in particular silk weaving, was a favored occupation of a group of Huguenots who had fled religious persecution in France to come to London in the late seventeenth century (Stewart and Weindling 1995). In 1717, this group founded the Spitalfields Mathematical Society, the forerunner of the London Mathematical Society, the national organization for mathematics in the UK. In marketing silk textiles, it was necessary to know the fineness of the skeins of a certain length, which were sorted according to how many skeins would weigh a pound. Doing this by eye was only possible after long experience, particularly when the skeins were fine. Euler was on the Saint Petersburg Academy's committee for weights and measures in 1737 and so took an interest in the problem. His work, published nine years later, concerned the most profitable application of simple and composite machines. The sensitivity of the equal-arm balance can be measured by the angle the beam moves through when a weight is added to one of the arms. He demonstrated that sensitivity is proportional to length of the arm and inversely proportional to it mass and distance between the center of gravity and the center of rotation.

Johann Heinrich Lambert considered a new form of balance, the bent lever, in 1758. This has the advantage that only the load is required, no other weights. The weight of the balance itself acts as the counterweight at an angle to the vertical less than that of the load. The sensitivity is measured by the rate of change of the angle when the load is added. If this is plotted against the load there will be a maximum, which is where the balance is the most sensitive. Finding the position of maximum sensitivity of a balance was a problem ideally suited to calculus.

In 1765, William Ludlam wrote in the Philosophical Transactions applying the calculus, or "fluxions," to produce a balance which is most sensitive at the point where precision is most required in the textile industry. Ludlam also proposed a bent-arm balance, using geometrical arguments and then applying fluxions to find the point at which the sensitivity was the greatest. This was done in the usual way by setting the second differential, the fluxion of the sensitivity, to zero. The balance could be arranged so that this point represented a value on the fine end of the spectrum, which was difficult to assess otherwise. Balances based on Lambert's and Ludlam's work were made by leading instrument makers in a rare example of a successful application of the new calculus (Biggs 2011).

As Europe moved from an agricultural to an industrial society, the requirements of mobility increased. The products of trade needed to be transported across the various countries, which was undertaken both by road and increasingly by systems of inland waterways. "Turnpike mania" is a term given to the period 1751 to 1772 in Britain, when the demand for road improvement was expressed in a number of Acts of Parliament. These concerns were reflected in the subject matter of those who fashioned themselves as "experimental philosophers." Desaguliers was the experimenter at the Royal Society, and was instrumental in bringing Newtonianism into Britain in the early eighteenth century. He was respectful of mathematics, but realized that mathematical theory did not always conform to the physical world. While machines with very little friction could be constructed to follow mathematical theory, weight-bearing carriages must have more friction. The design of carts and carriages was of interest, and from the early eighteenth century the lecturers on experimental philosophy played their part in this discussion by including experiments on wheel carriages in their courses.

Several Acts were introduced in Britain addressing various aspects of the design of vehicles. The experimental philosophers promoted mathematically superior high wheels of large radius. For example, James Ferguson, writing in 1771, likened the spokes of a wheel to a lever and suggested they should be as long as possible. However "low" wheels were cheaper, and allowed drivers to get up and down more easily. The most contentious subject was the widths of the wheel rims. In 1745, the width of wheels formed part of the pricing structure on the turnpikes for the first time: if the wheels were 8 inches or more in width they were exempt from restrictions on horse numbers as they were effectively improving the roads. In 1753, The Broad Wheel Act laid more emphasis on wheel widths, with 9 inches or more being unrestricted for weight, and more severe penalties for offences. By 1773, 9 inch wheels were referred to as "common." However, while wide wheels became the norm, high ones, or those with large diameters, did not. While the lecturers sought to legitimize mathematics, that legitimacy was not sufficient in itself to overcome commercial considerations, and the convergence of practice and mathematics was thwarted in this instance (Wess 1995).

The poor state of the roads was a major factor in the drive to provide an alternative form of inland transport. From the late 1750s in England, canals became a crucial method of moving goods, able to carry large quantities of raw materials and fragile finished products, both impossible by road. Britain was late in this development, as canals and locks were already a feature of the landscape in Italy, Germany, and The Netherlands during the seventeenth century. France had undertaken large-scale projects cumulating in the vast Languedoc canal which was completed in 1792 and "excited the admiration of the World" (Skempton 1996: 29). The first canal proper in England was funded by the third Duke of Bridgewater, who had been influenced by seeing the Canal de Languedoc in his youth. Because of the ease of movement the canal reduced the price of coal by a half (Conder 2017).

A small number of engineers including James Brindley and John Smeaton were responsible for the majority of the canals in Britain. The costings are easier to find than the engineering calculations. Smeaton and Brindley contrived an experiment to find the flow of water through a notch for various heads of water, but there is little evidence of mathematics beyond the very basic (Hadfield 1997). On three days in September 1768 in Edinburgh three engineers: James Brindley, Thomas Yeoman, and John Golborne, were asked a series of 21 questions on the proposed waterway between the Forth and the Clyde in Scotland. They were asked was it possible to provide a sufficient supply of water to carry 100,000, 200,000, or 300,000 tons of merchandise from Forth to Clyde and back, and at what cost. They were asked to advise on dimensions if they considered a width of 14 feet too hazardous. The answers were cautious and hedged with caveats. As for land surveying in general, canal construction involved a new profession whose members were familiar with basic mathematics and measurement, but no recent mathematics of the previous century was employed.

Hydraulic Engineering

Fountains and "water art" were popular with the wealthy and were increasingly devised using mathematical principles. It is well-known that Euler was blamed by Frederick the Great, King of Prussia for his fountains failing. "My mill was constructed mathematically. Vanity of Mathematics!" Frederick is reported to have said of his fountains at Sanssouci near Potsdam in a letter to Voltaire in 1778. However, failings in the wooden pipes and the inadequacy of the budget appear to have been the real obstacles in this case (Eckert 2002).

More important, though less spectacular, was the need for piped water for growing towns and cities. Piped water was incorporated into the vast subject area of fluid mechanics for the mathematicians of the period, discussed in detail in book two of Isaac Newton's *Principia* in 1687. In the second edition of 1713, Newton had to reconsider some of his results in fluid mechanics as the topic was so difficult and he was struggling to align theory and practice. (Smith 2001: 249). The extent of fluid mechanics can be seen in this chapter: piped water, wind and water power, ship design, and ballistics, all necessitated its study. It was hoped to provide all these various activities with an enlightened, unified, rational system of knowledge. (Darrigol 2005: vi) However, it has been remarked that even in the nineteenth century *hydrodynamics*, the mathematical analysis of ideal fluids, was a separate discipline from *hydraulics*, which was concerned with real situations, and the results could contradict one another (Eckert 2005: ix).

The mathematical relationship between friction and loss of "head" or pressure drop in the pipe was particularly sought after. Following Edme Mariotte who had addressed the problem in 1686, hydraulic engineers assumed that friction between running water and the walls of the container was proportional to the surface area in contact, and increased in some proportion greater than the velocity of the water. Investigations on the behavior of water in pipes were carried out in again in France by Claude Couplet, father of Pierre, who was the designer of the water system at Versailles in the early eighteenth century. It was found that friction was considerable in long pipes of small bores (Skempton 1996: 52). Dominico Guglielmini, an associate of the Italian school of fluid mechanics which was concerned with lagoons, flooding, and canals, had begun experiments on water flow in the 1680s, and by the 1730s he had introduced a form of viscosity, another complicating factor in fluid mechanics (Levi 1995: 209).

In a similar manner to that of bridge building, the analysis moved from the static to the dynamic. In 1727, Daniel Bernoulli, son of Johann in this large family of mathematicians, wrote a treatise on flow in pipes where he assumed the "vis viva" or mass times velocity squared (mv^2) was conserved. He elaborated on this in 1738 in his masterwork *Hydrodynamica*, where he extended the concept of wall pressure to moving fluids, arguing the force on the wall of a vessel consisted of static and dynamic components, although the dynamic part he addressed was only 1-dimensional. While Bernoulli's formula could only apply in very restricted circumstances, it was a new departure as pressure had previously been seen as merely a static property. This represented the first attempt to analyze the pressure of a fluid in motion. Both Daniel and his father Johann, who published slightly later, were therefore in part responsible for the "Bernoulli equation," demonstrating that pressure (not friction) actually decreases with velocity (see also Chapter 5).

In the mid-eighteenth century, fluid dynamics was developed considerably by Jean le Rond d'Alembert. In 1752, he departed from the Newtonian view of force as mass times acceleration (see Chapter 5 for a more in depth treatment) by considering the principle of equilibrium balancing pressure gradient, gravity, and inertial force at any point. Euler, who was frequently inspired by the work of others to tackle an extraordinary range of topics, chose the Newtonian approach when he considered the friction coefficient in pipe flow shortly afterwards. He argued that the friction of water in channels and that of solid bodies follow similar laws, and gave a formula for the expression for the loss of head as a result of friction.

As happened in many instances, Euler generalized the work of others, in particular the Bernoullis, to produce a coherent and groundbreaking theory. In a series of papers published in 1757, by considering the fluid as a continuum and using Newton's second law applied to infinitesimal elements, he succeeded in reducing hydrostatics to differential equations (Euler 1757: E225, E226, and E227). He then went on to consider the dynamical equations for a moving fluid, again proposing a general theory, which provided the basis for fluid mechanics in future. However, actually solving these equations was difficult and not achieved at the time. Crucially, Euler could not "conclude from his results a single new result fit for comparison with experiment" (Truesdell 1968: 149).

Smeaton worked on water supply in Britain and in 1780 he reported on the water service of the City of Edinburgh. However, there is little evidence that the mathematics of the French savants, the Bernoullis, or Euler had a significant impact on civil engineering in Britain. In the 1780s, the topic was revisited by Charles Bossut and Pierre du Buat, the latter suggested cohesion between the water particles and used the modern concept of viscosity. Du Buat balanced pressure gradient against frictional retarding force for a uniform flow, proving fluid friction did not depend on pressure. This thorny question of the dependency of friction on velocity was reconsidered in the early nineteenth century with a formulation containing quadratic and linear terms. In the meantime, hydrodynamicists could contribute little to the mundane issues of providing piped water (Darrigol 2005: 101).

Power

The issue of power was central to industrialization in Europe and became a prominent feature of the concerns of rational mechanics during this period. There was an impetus to measure and compare all available forms of power, the issue having been explicated by Christian Huygens as early as 1669. Huygens felt confident that comparisons could be made between the power of mills and that of horses or men. De la Hire and several others performed similar comparative experiments on various forms of power. Following in their footsteps, Desaguliers in 1744 advocated ascertaining how much a horse could draw before any mill or machine was constructed. Raising water or weights was becoming the standard way to compare power sources.

The largest number of urban affairs brought to the attention of the Royal Academy of Sciences in Paris during the eighteenth century was related to hydraulic phenomena and the improvement of water mills (Hahn 1971: 120). Steam engines were not as yet making a significant contribution to the industrial revolution. In the Paris Academy around 1700, there was a flurry of interest in the mathematization of the water wheel, and during the next few decades wind and water power were introduced into the Newtonian corpus, not just as abstract considerations concerning the vis viva controversy, but as specific instances of applied Newtonian mechanics.

The most influential writer for this period was Antoine Parent, a prolific mathematician and engineer working in Paris who wrote a series of papers on the topic that made use of the new calculus (see also Chapter 6 in Volume 3). His use of mathematics and the calculus followed logically if the basic premises were considered robust (Reynolds 1983: 202), but exactly these premises and

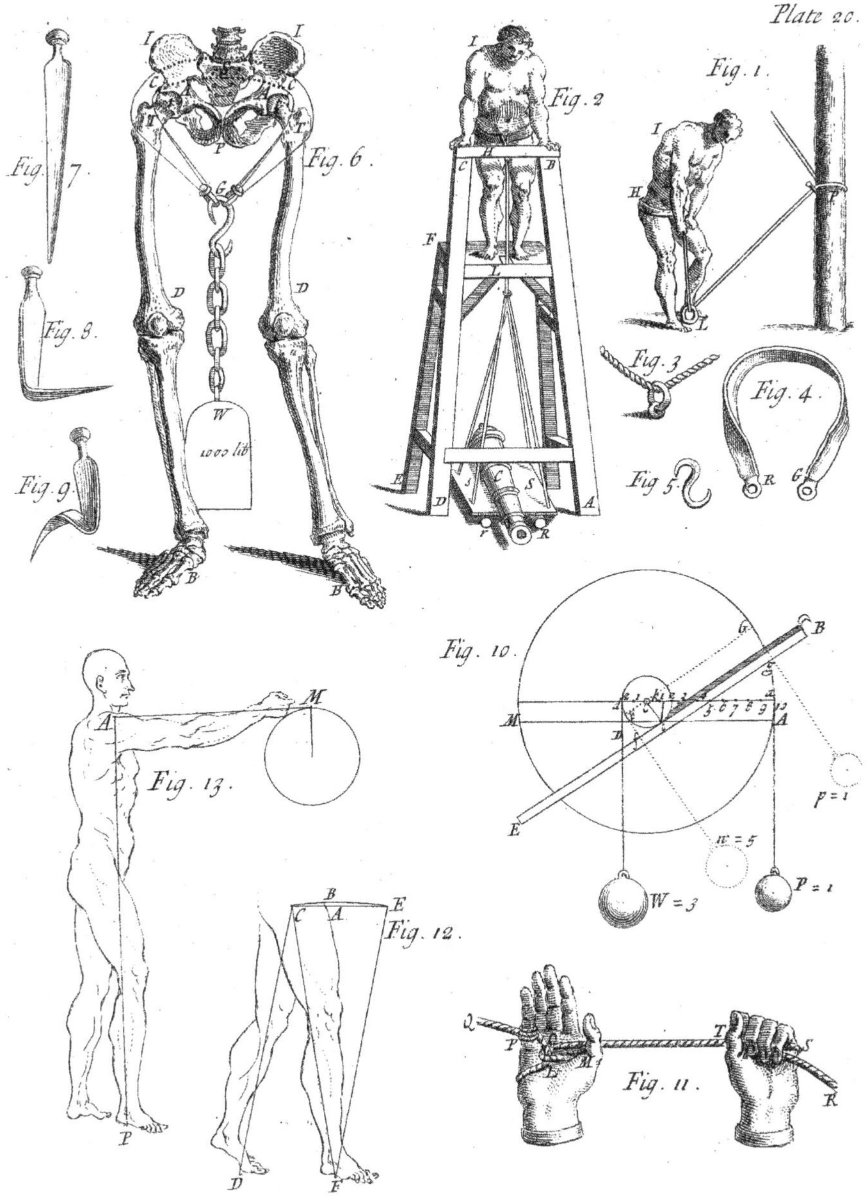

FIGURE 6.3. Plate 20 from J. T. Desaguliers, *A Course of Experimental Philosophy*, vol. 1 (London: W. Innys, 1745). Photo courtesy of Scientific and Technical Information Center / Internet Archive.

the empirical data he used would often be challenged by the next generation of engineers. Parent used a number of concepts to formulate his equation relating the water pressure before and after impact with the wheel blades. One such concept was the quantity obtained by multiplying area with velocity, introduced earlier by Mariotte who used blades of grass in a river and a seconds pendulum to measure this quantity. They further included the "efflux law," relating velocity of water spouting from the base of a container with the height of the fluid, and the "impulse law," relating pressure to velocity, and lastly the "spouting formula" relating pressure to the weight of the fluid. After some straightforward symbolic manipulation, Parent was able to differentiate his expression and obtain a quadratic equation. Setting this to zero provided a maximum value for the speed of the blades as a proportion of the original speed of the flow. If the first equation was correct, then it followed that the maximum efficiency of a water mill was when the wheel moves at a third of the speed of the river or channel. A similar process could be used to find that a maximum p representing a load of only four ninths of the load, which held the wheel static. Thus, the resulting maximum efficiency was predicted to be only about 15 percent. Parent was surprised by his result but it was often praised by later authors. Bernard Forest de Belidor, a leading French engineer, would call it one of the most important discoveries in the history of mechanics.

However, Parent had neglected the force of gravity and so had dismissed overshot wheels, as had Belidor. (Reynolds 1983: 215) Desaguliers had seen overshot wheels in action and asserted that they used ten times less water. This meant that theory and practice could not be aligned. In the Preface to the second volume of his *Course of Experimental Philosophy* which appeared in 1744, Desaguliers claimed he had studied water engines for many years, and that his friends had been expecting him to produce an analysis based on mathematical principles. However, he admitted that he had found the topic too abstract and there were no rules, which could be relied on.

In 1750, Smeaton approached the matter by using models, which he advocated as being the best means of ascertaining the results of his mechanical enquiries. The model he produced could be made to simulate an undershot, overshot or a breast mill. It was used to lift a weight in the scale pan while the water level in the top cistern was kept constant. Trials were made with different heights of water, rates of flow, and loads. Also the number of float boards could be changed. Smeaton was surprised that his results were different from those of Parent, who had become the accepted source, concerning the optimum velocity of the water, and complained that theory and practice did not correlate.

While wind power was not so central economically to industrialization, the mathematicians in the generation following Newton addressed it together with water power, and approached the issues in a similar manner. The workings of windmills are less intuitive than water wheels because the component that turns

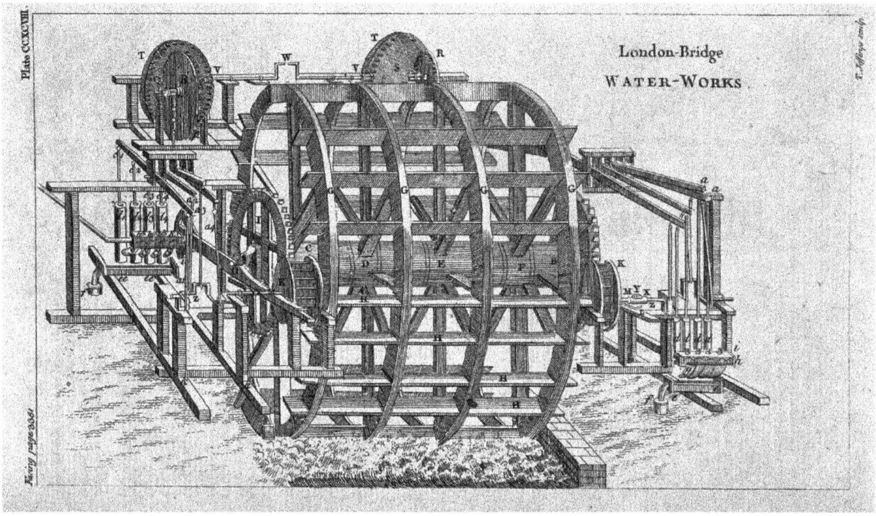

FIGURE 6.4. The Waterworks at London Bridge from *The Gentleman's Magazine*, 1749. Credit: The large waterwheel at the waterworks on London Bridge. Engraving by T. Jefferys. Wellcome Collection. Public Domain Mark.

the sail is at right angles to it. The other component of the force has to be resisted by the strength of the windmill building. There is therefore an optimum angle for the sail, which lies between no movement and destructive turbulence, and the mathematical procedure is similar for that of finding Parent's optimum velocity in the case of the water wheel. This question formed the focus of attention together with the angle of "weather"—the angle between the base and extremity of the sail.

Mariotte advocated the square law connecting velocity of wind and force on a sail in 1673. Together with de la Hire who published in 1699 and 1700, he set the scene for Parent's work. Mariotte had advocated an optimum angle for windmill sails to be inclined to the wind of sixty degrees. In 1701, Parent applied the same mathematics as he applied to the water wheel by taking the "differentiale," and set it to zero to obtain the maximum and minimum, arriving at 54.5 degrees for the maximum value of the angle of inclination of windmill sails. The Scottish mathematician Colin MacLaurin disagreed, claiming in his *Treatise on Fluxions* in 1742 that 54.5 degrees was only the optimum angle before motion had begun.

Smeaton also considered windmills and made a corresponding model using an artificial wind. He tested various inclinations and reached the conclusion that 55 degrees was actually the least effective. Instead, he advocated from 72 to 75 degrees, which he noted was the angle commonly used by practitioners. He recognized that there must be an optimum angle, but believed it could be

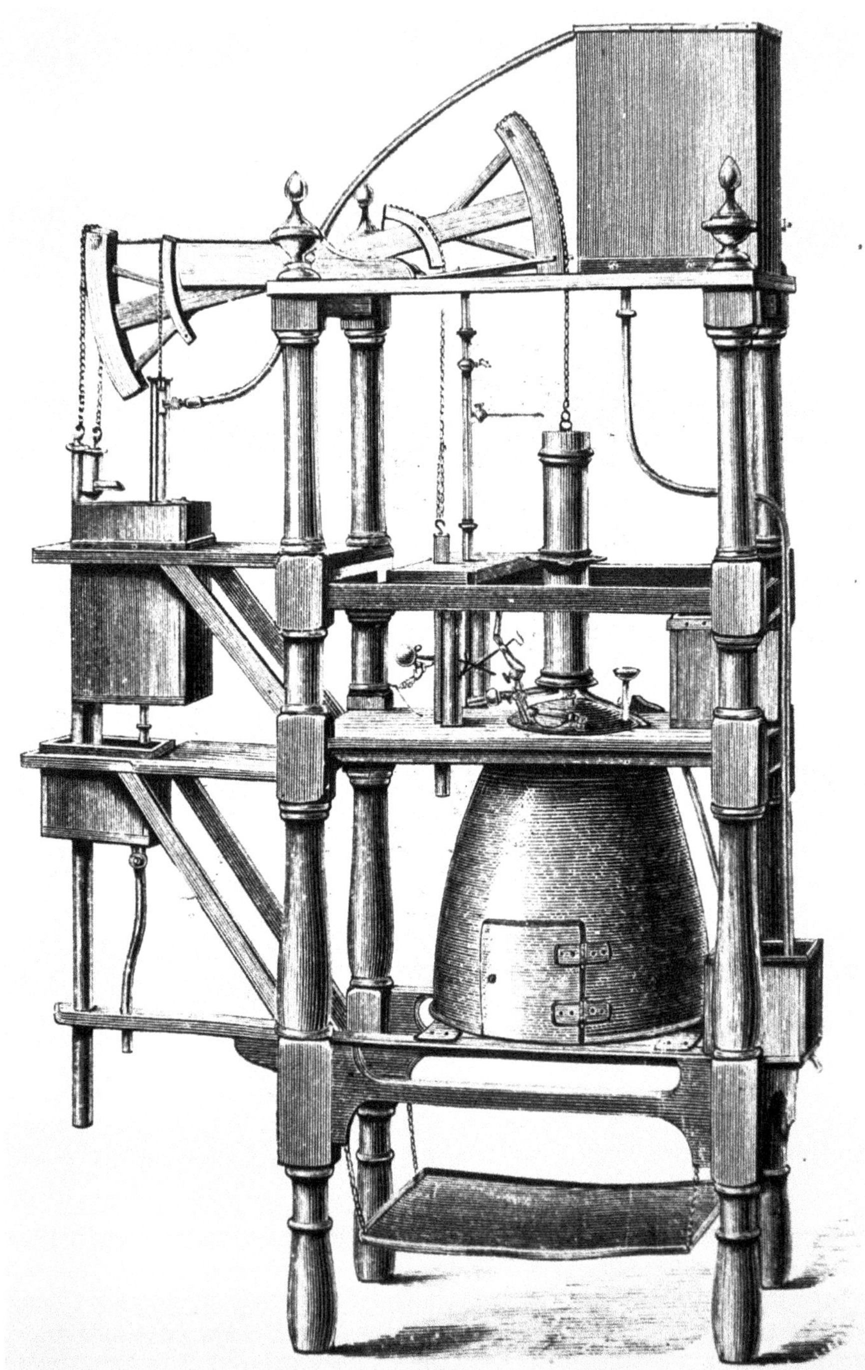

FIGURE 6.5. Engraving depicting a Newcomen steam engine. Photo by Universal History Archive / Universal Images Group / Getty Images.

found most easily through experiment. The practice of millwrights, as studied from examples in the Netherlands, appeared to make the best use of the wind without resort to mathematics (Davids 2008: 508–9).

As regards steam engines, the type of atmospheric engine based on the design of Thomas Newcomen was the only one to have a significant impact during this period. From 1712 they started to be used in Britain, and by the 1720s there were Newcomen engines all over Europe. The Newcomen engine became an important standard of efficiency at the end of the eighteenth century when Matthew Boulton and James Watt based their charges on the coal saved by using their fledgling steam engine rather than the established Newcomen one. Watt, in measuring the outputs of his steam engine proper, chose the variable horse as a direct measure of power, hence "horsepower," which became the industry standard, replacing measurements based on lifting water.

Initially Boulton and Watt engines were reciprocal, i.e., they worked a beam which moved up and down, which was fine for raising water from mines, but for little else. A rotative engine would have considerably wider application. In order to produce this Watt invented a sun and planet mechanism in 1781, and then a parallel motion device in 1784, the latter using the geometry of a pantograph. The conical pendulum centrifugal governor, which regulated the motion, was introduced in 1787. Indicators involving mercury barometers were used to record the pressure in the cylinder from the 1790s. In 1796, John Southern, an engineer working with Watt, suggested replacing the pointer by a pencil which could trace the line on card to form a closed loop. This new pictorial approach, with obvious potential for integration, contributed to new avenues in mathematics in the following century.

THE REQUIREMENTS OF IMPERIALISM

Taxation and Gauging

Turning next to the requirements of imperialism, much of Europe was involved in a series of wars during the period under discussion. Most notable were chronologically the Great Turkish War, the Nine Years' War, the War of the Spanish Succession, the War of the Austrian Succession, the Seven Years' War, and the French Revolutionary Wars at the end of the period. This increased level of conflict necessitated fund raising, which was done largely through direct taxation or the "excise." Excise tax is a duty on manufactured goods, levied at the moment of manufacture on goods produced domestically.

Many commodities were subject to the tax, but one of prime importance that demanded knowledge of mathematics was alcohol. By 1714, excise accounted for over 40 percent of all tax collected in Britain; the largest single tax. By 1733, it accounted for over half of all tax. Unlike land tax it was administered centrally using an appointed body of crown employees which numbered 2800 by 1780 (Brewer 1989: 83). Another important product was sugar, which was becoming more in demand as tea-drinking increased. The product to be taxed in this case was molasses. In both cases, the determination of the volumes of barrels, known as gauging, was key (see also Chapter 6 in Volume 3 for more details on gauging).

The Excise became the biggest government department and had the most contact with the public. It is well known from the writing of Tom Paine who was an exciseman, that he studied for 14 months before becoming qualified in 1761. The gaugers walked miles or rode, with their equipment of a set of gauging rods and a slide rule. One of the first books "to instruct and assist his majesties officers for the excise especially supervisors and gaugers in the ready and faithful discharge of their duties" was Thomas Everard's *Stereometry or the Art of Gauging Made Easie by the Help of a New Sliding Rule* published in 1689. The work incorporated decimals, then multiplication and division using the slide rule, square roots, cube roots, how to find areas and volumes using gauge points in a similar manner to girt lines, cones, spheres, cylinders, rhomboids, and accounting. It had an appendix on conic sections which was necessary to distinguish various forms of casks or "tuns," and to recommend rules for finding the content. Calculating the volume was based on a choice of one of four shapes. Later these men had to study Cocker's *Arithmetic* and Lightbody's *Art of Gauging*. Excise men were the largest group of mathematically literate professionals during the period.

As in many other areas, the new mathematics of the calculus was also put to work here. In Britain, the prominent Scottish mathematician Colin MacLaurin was involved in the determination of the volumes of barrels (Grabiner 1998). He had produced beautiful mathematics and had identified four basic shapes of barrels. He then went on to consider the errors that arose from taking the frustra of different conic sections to be cylinders, errors which he noted were always to the disadvantage of the trader. He found that the differences remained the same for each shape no matter where on the conic section it was taken. The involvement of mathematics was helpful in assuring an impartial authority in areas of tension and sometimes even violence.

Navigation

Probably no other activity attracted so much mathematical resources as navigation (see also Chapter 5 in Volume 3). The stakes were high, especially for the imperial states. Ten thousand tons of shipping passed through Calcutta each year between 1715 and 1725, and the British and Dutch East India Companies were the size of nation states. By 1786, the French had a fleet of 5268 merchant ships with a tonnage of 729,340 in order to supply Europe with sugar, coffee, tea, and tobacco (Landes 1983: 144 and 159). European states were in competition for trade and strategic interests across the Indian Ocean and the Atlantic, driving the need for more reliable navigation, and in particular the need to find longitude.

After a ship had left sight of land, the only method of finding position was by compass for direction and log line for distance, neither of which were very reliable (see Chapter 5 in Volume 3). Latitude was relatively easily found by the maximum height of the Sun at noon or, in the northern hemisphere, the Pole Star at midnight. However, longitude was a different matter. Half a dozen methods were in use, but by the end of the period two strategies, equally difficult to implement, had emerged as being feasible. One was to have a precise terrestrial clock to keep time from a primary meridian in order to compare it with local time, the other was to use the regular motions of the heavenly bodies to compute position. The thorny issue of longitude stimulated the pursuit of natural philosophy and the production of instruments as well as the founding of institutions such as observatories and schools. Spain and the Netherlands had offered prizes for finding longitude before Britain, and in France the Paris Observatory was founded in 1667 with the central aim of observation of celestial bodies for this purpose. Several institutions promoting mathematics were introduced just before or during this period in Britain: the Royal Mathematical School at Christ's Hospital in 1673, the national Royal Observatory in 1675. With the Longitude Act in 1714, a prize was set for those who would provide better and more precise means to find longitude at sea.

There is a long list of mathematicians across Europe who saw the potential of lunar theory for solving longitude. They could rely on the considerable resources of data from leading observatories. However, as the motion of the moon is a particular instance of the intractable three-body problem in mechanics, a mathematical accurate theory of the moon was by no means evident. The start of this "harrassment of the Moon," as William Whiston put it in 1728, was a visit by Newton to Greenwich in September 1694, when he was shown Flamsteed's lunar data. At that time, Newton felt there was just a possibility that the moon's motion could be understood in a way which would solve the longitude problem. Gregory related that Newton felt "the theory of the Moon is within his grasp" (Westfall 1980: 451).

Newton's *Theory of the Moon's Motion* appeared in a pamphlet in 1702, and it was included in a slightly different form in the second edition of the *Principia* in 1713, when a reviewer called attention to the longitude problem. However, when giving evidence in his statement to the British parliamentary commission on longitude in 1714, Newton listed the various possible methods before denigrating all of them. In his opinion the method based on the moon's motion was only exact enough to find longitude within two or three degrees, not within one degree. Lunar theory again appeared in the 1726 English edition but Newton still considered the motion of the moon was not yet sufficiently known to ascertain longitude with sufficient accuracy.

A breakthrough came in 1752 when Tobias Mayer, a German mathematician and astronomer, produced lunar tables that were sufficiently accurate for use in tables for navigation. Mayer's widow would later get a part of the prize money of the Longitude Act. The result of this long endeavor spanning three generations was the *Nautical Almanac*, which arrived in 1767, and the accompanying *Tables Requisite*, containing data that did not need recalculating on an annual basis. The *Nautical Almanac*, so it was claimed, enabled seamen to use the lunar-distance method. However, while 10,000 copies of the *Tables Requisite* were printed by 1781, this is no guarantee of use: 6,992 remained unsold by 1784. Ships' masters were supposed to get a certificate from the Portsmouth Royal Naval Academy, the idea being that when the ship docked there they went on a course on the lunar-distance method. However, older masters were refusing to learn, and by 1771, four years after the publication of the *Nautical Almanac*, only 14 had been instructed (Howse 1989: 86).

A survey of navigators' textbooks, instruments, and ships' logs in the 1770s has found no evidence of new methods being introduced generally as an immediate result of the publication of the *Nautical Almanac* (Wess 2015). Maskelyne's assertion that it demanded "only care in the computer, and [. . .] no particular knowledge of spherical trigonometry" is technically true but misleading. The computation of lunar distances was tedious, complicated and opaque compared to calculations based on dead reckoning, and without appropriate training remained beyond the capabilities of ordinary navigators. Benjamin Martin's preface to part 2 of *The Mariner's Mirror* dated 1782, clearly states that the lunar distance method for longitude was so difficult as to be of little use to the ordinary seaman.

Instruments also indicate a continued reliance on dead reckoning, that is, estimating the current position of a ship from a previous known position using heading, time and the ship's speed measured by a chip log. The Gunter rule, a device consisting of various trigonometric and logarithmic lines plotted on a boxwood or occasionally brass rule, was designed for navigators in the seventeenth century, based on the work of Edmund Gunter. These were still popular in the mid-eighteenth century and the problems it was designed to solve were exclusively those of dead reckoning. A study of ships' logs also suggests that in the Royal Navy, except on elite and specific voyages, the lunar distance method was not used until the first decade of the nineteenth century at the earliest. In the East India Company, a brief study suggests that only ships of exploration were using them until the 1790s, and then it was just as likely a chronometer would be preferred.

The other new method of finding longitude was by (terrestrial) clockwork. Knowing the time at a set starting meridian, and comparing it with the time locally could give longitude. While involving far less mathematics than astronomical methods, the topic of clocks for longitude also gave rise to calculation. Following considerable work on pendulum clocks, Huygens described a spring-driven mechanism in a letter to the *Journal des sçavans* in 1675. However, at that time springs were no easier to control than pendulums due to temperature variations (Yoder 1988:161). It took the painstaking work of John Harrison in the eighteenth century to develop a viable piece of clockwork leading to the production of "chronometers" by the end of the period. While this name denoted a clock for use at sea, it also acknowledged the difficulties of producing a time-keeper to function in this environment. In 1780 Joseph Banks and Alexander Dalrymple officially named the device, the term "time-keeper" being reserved for a perfect chronometer

Sophisticated mathematics did play some part in the construction of clocks for longitude. In 1695, de la Hire had suggested that the mathematically beautiful cycloids, where a point on a circle rolls on another circle (epicycloid), or along a straight line (cycloid), or inside a larger circle (hypocycloid), could be used to optimize the shape of gear teeth. They should move with uniform motion and roll without slipping. However, using calculus, Euler demonstrated in a publication of 1760 that it was not possible to have gears operating with uniform motion and no friction simultaneously (Euler 1760: E249). In fact, the epicycloidal gears were good for clocks because they kept time, but did not eliminate friction. The chronometer maker John Arnold took out a patent for epicycloidal gear teeth in 1782, but his design did not influence later makers as it was difficult to reproduce until wheel-cutting engines could be used.

Even by the end of the eighteenth century chronometers were not considered reliable. John McCluer, first lieutenant of the East India Company Marine at Bombay, was taking the time by chronometer every hour and altitudes with an octant in 1786. He complained that he had been sent two Arnold chronometers by the East India Company that were more trouble than useful. On a later voyage in 1789–90 McCluer took lunars as well as using a chronometer, but put little faith in his results. Chronometers were not mentioned in textbooks in Britain until the 1790s. This therefore constitutes another example of where the beautiful new mathematics of the seventeenth century was not readily applicable during the eighteenth century.

Ship Design

Wind was the form of propulsion for ships engaged in imperialist activities in the late seventeenth and eighteenth centuries. The European nation states, particularly France, were determined to obtain naval supremacy, so mathematical solutions to the problem of ship design were highly desirable. Referred to by Euler as "ship science," it involved the most difficult problems facing mathematicians of the time. It required a consideration of forces acting simultaneously on the hull in one fluid and the sails in another.

An important aspect of fluid mechanics in this case was the behavior of solids in fluids, especially their resistance. Newton considered resistance, arguing it was equal to the density of the fluid medium times diameter of the solid body squared times the velocity of the solid body squared or $r=\rho \times d^2 \times v^2$. He could not definitely establish what he expected, that resistance was proportional to the square of the velocity. He believed the particles of gases such as the air could be individualized, but those of liquids could not, therefore the theories had to be different. However, he was compelled to consider any fluid medium as consisting of particles which were not affected by each other, as the problem would have been too difficult otherwise. Mariotte had investigated simple shapes of solids in a river and in air, in order to measure their resistance, finding resistance was proportional both to velocity and to the density of the fluid medium. He used the impact theory in his work, published posthumously in 1696, which was lent authenticity by Newton the following year (Calero 2008: 15).

Paul Hoste, Jesuit Professor of Mathematics at Toulon, complained that ships were designed "almost entirely by the eye." Writing in 1697 that, "The art of constructing ships, which is so necessary to the state, is the least perfect of all the arts," and advocating a mathematical approach (Nowacki and Ferreiro 2011: 149). Hoste argued, without using calculus, that "the force to carry sail" or its stability, was "easily known" to be proportional to its weight times the distance between its center of gravity and its center of displacement. So, dubiously, the higher the center of gravity the more stable would be the ship. Sections of a ship were considered to be segments of circles with the center of gravity at the center of the segment. "Heel" or leaning, "trim" or shape, and "yaw," the propensity to turn around a vertical axis, all had to be considered.

As well as stability there were issues of propulsion to consider. The aspect of the wind on the sail was considered by Huygens just before he died, associating sail forces with a v^2 law in a Newtonian approach. Jacob Bernoulli used differential calculus to analyze curves or shapes of sails in wind, known as the "velaria" problem, in 1691 (Calero 2008: 160) Jacob defined clearly the apparent wind velocity, i.e., the vectorial difference between true wind and ship speed, as relevant for the sail force. He was led to a second order differential equation for the shape of a sail, giving rise to the "velaria" curve, in the same family of curves as the catenary. His brother Johann Bernoulli published a treatise characteristic of the state of the art at the time in 1714. He also applied Newton's impact theory of resistance, including both the hull resistance due to underwater hydrodynamic forces and the thrust acting on the sails due to the aerodynamic effects of wind. He achieved this in a mathematically valid way, just as Parent had done in the case of water wheels, but again the agreement of these predictions with empirical observation was poor. The French ministers of the navy, Jean-Baptiste Colbert and the Count de Maurepas, were keen to exploit mathematics to challenge the sheer scale of British sea power. When the Royal Academy of Sciences in Paris set the challenge to determine the optimal configuration, number, and height of masts for sailing vessel propulsion in 1727, the prize went to Pierre Bouguer. His later manual, *Traite de la Navire* (Treatise on Shipbuilding) of 1746 would cement his influence in the field.

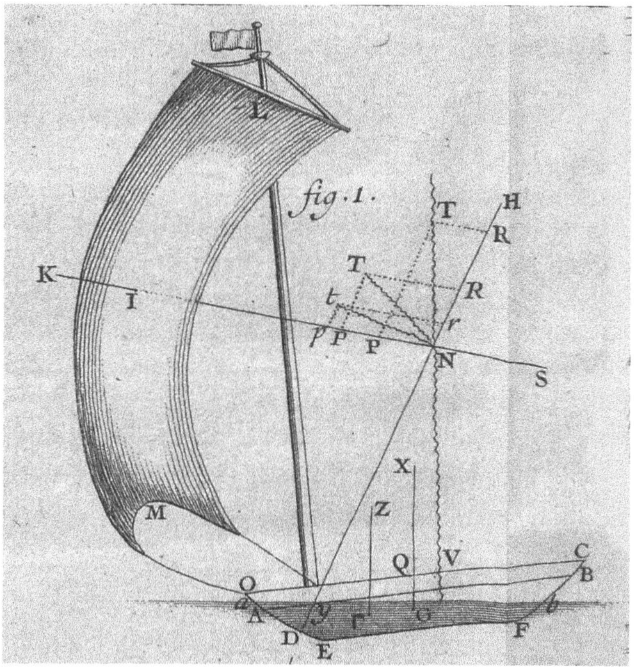

FIGURE 6.6. Hull resistance and sail forces intersecting in the "point vélique" in P. Bouguer's *De la mâture des vaisseaux* (1727). Photo courtesy of Jombert, ECHO, Max Planck Institute for the History of Science, Library, Public Domain.

Bouguer used the "point vélique" and incorporated Hoste's ideas (Ferreiro 2010). The "point velique" was defined as the point of intersection of the sum, or resultant, of the resistances of the hull, with the resultant of the forces acting on the sail. Bouguer assumed buoyancy and weight were equal and opposite, and pressure was normal to the surface of the submerged body. He used approximate shapes for the ship and took prism sections, integrating them using the trapezoid rule to find areas and volumes. He advocated a metacenter above which the center of gravity could not be found. Bouguer also included consideration of the shape of the stern, which had been ignored until then.

Considerations of balancing greatest velocity with safety led Bouguer to a series of statements regarding proportions relating the heights of masts, the length and breadth of the hull, and the area of the sails. He was extremely concerned with stability. The subject quickly found root in Spain, Sweden, France and Denmark but less so in Britain or The Netherlands. In France the École du génie maritime was set up in 1741 and in Spain Bouguer was influential through Juan y Santacilia's use of his formulas in his publication of 1771. Swedish ship science benefited from Frederik Henrik Chapman who took mathematics lessons from Robert Simpson in Woolwich and constructed ships using Bouguer's methods, publishing in 1775. However influential Bouguer became, he remained cautious, recognizing that there was too much dependence on abstract principles.

Euler's work on ship science began with the above-mentioned 1727 prize, but he was stimulated further in 1735 when he was asked to review a work by Cesar Marie de La Croix, general commissioner of the French fleet (Calero 2008: 150). This period of focus was between 1737 and 1749 during which time the state Academy in St. Petersburg commissioned *Scientia Navalis* which was published in 1749. Euler suggested that pressure of water was proportional to depth, that the weight of the vessel could be assumed to be at the center of gravity, and defined stability from the restoring moment. Using Leibnizian notation he arrived at a general 3-dimensional case for a floating body of arbitrary shape.

Although Euler usually used impact theory, in this particular case he turned to the difficult concept of streamlining, arriving at the paradoxical result that fluid resistance was zero. D'Alembert also turned away from impact theory. Writing in 1743 in *Traite de Dynamique*, he rejected forces in favor of masses and velocities, looking at more general systems. In 1749, d'Alembert submitted an essay on a new theory of resistance, which involved the cancellation of velocities. Again, he found that the total resistance on the body is zero, now known as d'Alembert's paradox even though Euler had actually formulated it first. They had reduced the physical problem to a mathematical one but could not solve the equations or equate their results with observation. Euler returned to ship science periodically throughout his career, for example on stability in 1759 (Euler 1771: E415), using principles of mechanics and integral calculus.

According to impact theory, the hull resistance acts obliquely upward. It was believed that masts and their sails should be arranged in such a way that the resultant sail force would have a horizontal forward component, which acted through the point vélique, and during motion at constant velocity this should be in equilibrium with the horizontal component of the resistance. However, these considerations on the point vélique were grossly misleading but also superfluous (Nowacki 2008). It was assumed that the sail area of several masts could be lumped in the centroid of all sail areas and that the resultant sail force would act through this point. Therefore, the predictions of sail forces were unrealistic in both magnitude and direction. Again, the mathematics was sound but the physical assumptions were flawed. Notwithstanding, these principles were taught in French naval institutions (Ferreiro 2007).

Euler's last book was his general treatise on ship design, written in 1773. Henry Watson, who was a chief engineer in Bengal, translated into English in 1790. Discussing the arrangement of sails, Euler claimed that for the ship to move in the desired direction most efficiently the moving force should act in the direction of travel. Therefore, the sails should be arranged perpendicular to that force, and perpendicular to the principal axes of the ship. However, this theory predicted that the more oblique the wind the less will be the velocity of the vessel, which is not the case in practice because the back sail obscures the others. Euler came to recognize that the principles of hydrostatics were not sufficiently developed to determine the pressure on the ship (Sandifer 2007: 273).

Towards the end of the period more work was done involving models. In France Jean-Charles De Borda and Charles Bossut used fluid streams, and in Sweden Fredrik Chapman performed towing experiments using models. In Britain, William Emerson, John Sewell, and George Atwood also used models to investigate friction, but then only crude geometrical shapes were investigated (Schaffer 2004). Jorge Juan's work of 1771 looked for an alternative to impulse theory which, although largely ignored at the time, showed the way for future research (Calero 2008: 41). There are widely diverging assessments of the value of these works among historians. While some argue that scientific principles were applied during the eighteenth century, much evidence is available to argue that all wooden ships were based on a previous wooden ship until the end of the period (McGee 1999).

Ballistics

Like ship science, ballistics was important to eighteenth century states that were in competition for authority over the oceans. As William Mountaine, who published a book in 1744 on fighting ships including the use of great guns, stated in 1781, mathematicians needed to be involved in gunnery because competing nations could not remain in a state of peace indefinitely. And like ship science, ballistics involved resistance in fluids. Newton had been ambivalent regarding the relationship between velocity and resistance. In the third edition of the Principia, the English version of which appeared in 1729, the propositions at the start of Book 2 assume a resistance in proportion to the velocity. However, in the following Scholium Newton made it clear that this is only a mathematical hypothesis, and went on to consider the case where the resistance was proportional to the square of the velocity. Nevertheless, writers of textbooks on gunnery in the early eighteenth century overlooked these reservations and quoted Galileo, Torricelli, Halley and Cotes, all of whom had neglected fluid resistance completely (see also Chapter 5 in Volume 3).

Sébastien Le Prestre de Vauban, the engineer general of France, produced a text aimed at military personal and gentlemen which was reprinted many times in the late seventeenth and early eighteenth century. He asserted that cannon balls travel in a parabolic line as proven by Galileo and Halley, so tables of elevation versus range gave 45 degrees as the optimum angle. John Grey's *Treatise on Gunnery* published in 1731 was similar. Grey started with the premise that spaces run through are in proportion to their times and velocities, resulting in a parabolic curve. He set questions on which angle the cannon

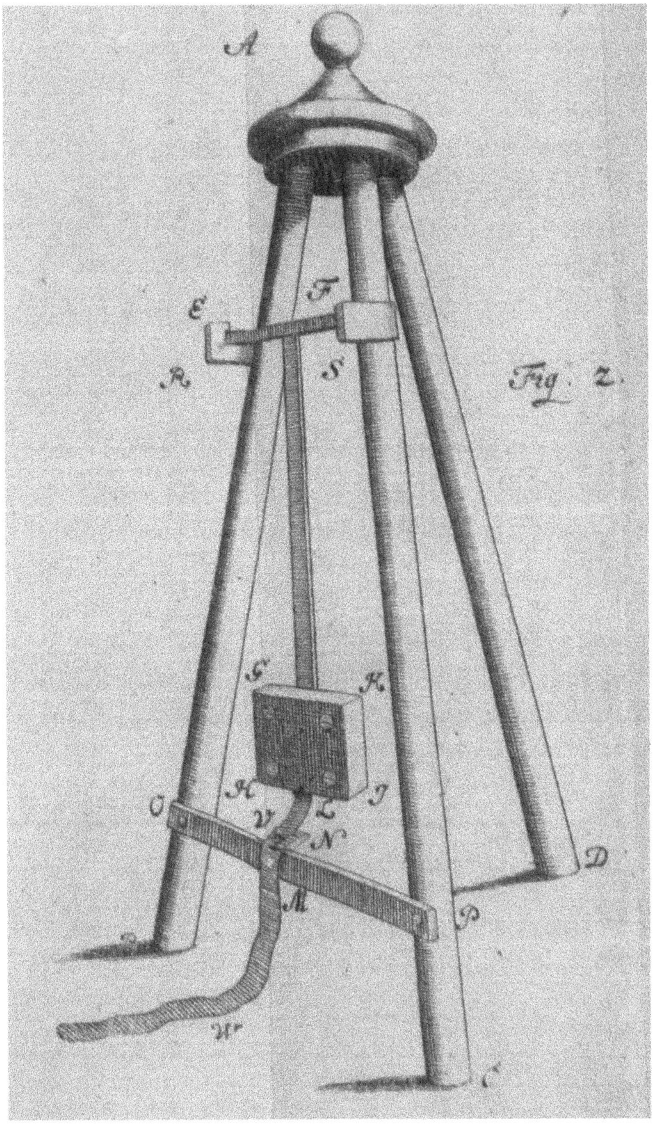

FIGURE 6.7. Robin's ballistic pendulum for determining the force of cannon balls. Photo courtesy of ETH Library Zurich, Rar 4648, Public Domain.

should be set, what horizontal distance the ball will travel, what force it will be given, and to what height it will rise using logarithms in what becomes complicated but not advanced mathematics.

The mold needed to be broken and this was done by Benjamin Robins. Keen to be appointed as mathematical master at Woolwich, he started to write his book on fortification and gunnery, the first edition of which was published in

1742. Robins was the only writer for a considerable time to challenge the acceptance of the parabola, recounting that he had confirmed by experiment that it did not solve the problem. Instead he found very considerable resistance at high velocities. He expected his results would be contested because they went against the prevalent views. Newton's ambivalence had been ignored. Robins assumed that resistance was proportional to velocity squared approximately. However, he recorded the empirical results of his experiments and did not go into the calculus.

Robins' illuminating ideas were not taken up immediately, but fortunately they received attention from Euler, who early on had recognized the need for sublime mathematics in the calculation of flight paths of cannon balls (Sandifer 2007: 383). Euler's interest in ballistics began in 1727, his first year at the Saint Petersburg Academy where he had been present when Daniel Bernoulli conducted several experiments involving cannons. Daniel published his results and then described them again in his *Hydrodynamica*. Euler was interested in Bernoulli's work and later Robins' work, to the extent that he wrote an extensive preface for a new edition of Robins in 1745. In this preface, Euler advocated the use of calculus strongly, arguing it was in the public interest to understand the motion of projectiles.

Euler wrote again in 1753 on "Research on the true curve that is described by bodies shot through the air or in any other fluid" (Euler 1755: E217), finding that there was theoretical and experimental evidence that air resistance should be proportional to the square of the body's speed through the air. Euler divided his discussion into two "branches" of the curve, the ascending branch, and the descending branch. By applying a constantly decreasing horizontal component of velocity, he demonstrated the curve was not a parabola. He also showed that the point at which the curve had its greatest curvature was not at the vertex, but was a point on the descending branch near the highest point. He also showed that the point where the velocity was minimum was not at the highest point either, but at a point beyond this. This curve could not be symmetric. Euler finished this paper with some tables that show how to find the true curve accurately and how to find various things of interest to artillerymen, like the height of the vertex and the speed of the projectile at various points along its trajectory.

There appears to have been little immediate impact of this work in Britain for more than thirty years, but in the 1770s a more mathematical approach was urged, and Euler and Robins began to be cited, for example by Hugh Brown in 1777. Brown supported the use of "sublime mathematics," i.e., the calculus, and an optimum angle of less than 45 degrees, and was concerned that the topic had not yet been fully mastered. Commenting on Robins' assertion that the track described by the ball was not a parabola, Brown complained that nothing had been done since as the problem was so difficult. There was a need for fluxions.

In the same year, Alexander Ewing also cited Euler, Robins, and Brown in *The True Principles of Gunnery*. Ewing quoted Euler's opinion that mathematics was essential for discovering the true path of the projectile. The optimum angle for the start of motion would be less than 45 degrees if air resistance was taken into account. Other countries, too, would start employ Euler's approach and thus use differential calculus to compute ballistic tables, such as Henning Friedrich von Graevenitz in Prussia (1764) or Jean-Charles Borda in France (1772).

When the French modernized their artillery education, the minister Turgot suggested to Louis XVI in 1774 that two works by Euler were purchased; the construction and maneuver of vessels, and the commentary on Robins. The professor Jean-Louis Lombard would later translate them into French for the cadets of navy and artillery schools. It has been suggested that Lombard's student, Napoleon Bonaparte, who himself wrote a treatise on gunnery in 1788, owed part of his later military success to this mathematical intervention.

CONCLUSION

The mathematicians of the seventeenth century, and particularly Newton, raised the expectations of mathematics for solving a range of problems facing industrializing and imperialist nation states. The Enlightenment belief in rationality, reason, and order, was in tune with a desire to mathematize.

The requirements of these nation states in this period encouraged an increasing mathematical literacy of substantial sections of the European populations. Often using slide rules and equations as recipes, navigators, surveyors, excise men, and even weavers, were nevertheless using mathematics for finding where they were, how much land they were dealing with, how much liquor was in a barrel, or what the value was of the silk they had produced.

More advanced mathematics was used to provide rules, which it was hoped, would give optimal and completely general and dependable solutions everywhere in all situations. In particular the calculus provided a method of finding rates of change of one variable with respect to another, and rates of change of rates of change, with the resulting maxima and minima providing mathematical formulations for efficiency in many instances. However, real physical situations were frequently too complex to be understood, so the beautiful mathematics applied, although faultless in itself, gave rise to meaningless and sometimes plainly wrong results.

Many of the difficulties with the application attempted in the eighteenth century were resolved in the nineteenth century, but this period saw the introduction of a general acceptance that mathematics held the key to understanding, even if it fell short in many technologies at the time. As J. H. Lambert pointed out in his acceptance speech at the Berlin Academy (1765), the mathematical sciences and experimental physics needed each other, but:

> the difficulties that the physicist and the geometer encounter, accumulate when combined, therefore it is necessary to neglect neither computation nor experiment. [. . .] Computation brings precision and universality. Experiment verifies both and discovers each omitted or falsely admitted circumstance. By neglecting computation and theory that is its basis, one makes experiments at random and without design. By neglecting experiment, one runs the risk of creating chimera and produce calculation applicable in whatever other World but the World we live in. (Lambert 1765: 508–9)

Lambert warned both for badly made experiments that "mar the progress of good theory" and for the proper love invested in hypothetical theories and calculations that made mathematicians averse to testing their theories. This sentiment sums up the relationship of mathematics to technology in the period. If, according to Morris Kline, "Mathematics is valuable largely because of its contributions to the understanding and mastery of the physical world" (Kline 1960: viii), then, in the eighteenth century, "Man's finest creation for the investigation of nature" was not yet in a position to help out everywhere.

CHAPTER SEVEN

Representing Mathematics

Images of Mathematics in Philosophy and Literature

FRANZISKA BOMSKI

Mathematicians were not the only ones interested in mathematics in the eighteenth century. It also exerted a great fascination on other professions. Mathematicians as well as non-mathematicians attributed certain transcendental, epistemic, aesthetic or ethical properties to mathematics and thus constructed images that—unlike the concrete practice accessible only to experts—could have a broader cultural impact. For example, the emergence of infinitesimal calculus in the seventeenth century marked the arrival of a powerful problem-solving tool whose abstract language came to have far-reaching applications in the natural sciences. So seemingly limitless was its potential that quite a few Enlightenment thinkers wondered whether it might serve as a universal calculator, capable of computing new knowledge in disciplines that also lay beyond the natural sciences such as metaphysics, ethics, and aesthetics (see also the Introduction). The ways in which philosophers, writers, and artists in the eighteenth century engaged with mathematics are numerous and diverse. They referred to various mathematical developments, interpreted them in many ways and drew very different images of the mathematical, some of them colorful, some of them horrific.

In the following, important aspects of these images of mathematics will be highlighted, without claiming to exhaust the subject. This chapter thus deals with the cultural history of mathematics in a very specific but nonetheless very important sense. Mathematical concepts in themselves play a merely ancillary role. Instead the focus is on a selection of the cultural representations of mathematics in the eighteenth century, many of them conceived by mathematical laymen who nevertheless noticed the growing importance of mathematics in scientific advances and explored its connection with changing worldviews.

The development of the history of ideas, even if it deals with mathematics, does not strictly adhere to the caesuras marked by dates. Therefore, we will engage in some retrospection into the late seventeenth century, as well as looking forward into the early nineteenth century in order to trace important traditions in the history of ideas. This essay is roughly divided into two main sections. The first part deals with philosophical images of mathematics conceived by Gottfried Wilhelm Leibniz, Jean le Rond d'Alembert, Moses Mendelssohn, Immanuel Kant, and others. Their reflections on mathematics pursued the question of the nature of mathematical truth in general, its position in the order of knowledge and, closely related, its potential as a leading discipline and a universal role model for epistemic progress (compare also with Chapter 4 in Volume 3).

The question of the relationship between the philosophical view of nature and the mathematical description of the cosmos assumed a central place in the poetical thinking of German Romanticism in the late eighteenth and early nineteenth centuries. Rejecting more rigorous forms of argumentation in favor of aesthetic knowledge and sensory intuition, German Romantics probed new perspectives. They embraced literature as a means of philosophizing, while playing with semantic ambiguity and employing metaphors along with other forms of associative thinking. The second part of this essay will provide insights into poetic reflections on mathematics using the examples of Friedrich Schlegel, Novalis, and Johann Wolfgang Goethe. Like the philosophers mentioned earlier, they have been chosen not only because they are important representatives of the literary and intellectual life in the eighteenth century, but also because each exemplifies how images of mathematics are woven into their literary work.

A brief explanation of the term "image" as it is used in cultural-historical research on mathematics may be useful. The historian of mathematics Leo Corry uses the term *images* of mathematics to analytically separate "questions about the subject matter of the discipline" which he calls the *body* of knowledge from "questions about the discipline *qua* discipline, or second-order questions" which he calls *images* of knowledge:

> It is the aim of a discipline to answer the questions of the first sort, but usually not to answer questions of the second sort. These second-order questions concern the methodology, philosophy, history, or sociology of the discipline and are usually addressed by ancillary disciplines. (Corry 1989: 411)

Corry's terminology roughly corresponds with Cathryn Carson's distinction "between interpretation, understood as explicating the mathematical structure of a theory in its own precisely defined domain, and implications, in the sense of defining the significance of the theory for broader realms of experience" (Carson 1995: 646, ann. 11).[1]

Such *images* of mathematics can be characterized by three main features. First, and most importantly, they are claims about mathematics that can only be understood appropriately by considering their historical contexts and the social environments of their creators. Second, and closely related, they are "things that people construct" to establish alliances between mathematics and other realms of knowledge: "contingent creations, reflecting as much about their creators as about the domains under discussion" (Carson 1995: 646f.). Therefore, there can be multifarious *images* of mathematics, they can complement or contradict each other, some of them can stay relatively stable over a long period of time, others may drastically change in even a short span of time or between different social groups in the same era. Third, there are different levels of interaction between the *body* and the *images* of mathematics. Both Corry and Carson point out that certain *images* held by individual scientists or by a broader scientific community or even society in general combine to be responsible for shaping the *body* of knowledge by determining which theories and problems are worth further research (and funding) and which are not. However, for a cultural history of mathematics, the *images* are of interest, even when there is no such direct influence on the mathematical *body* of knowledge: the *images* of mathematics are one important element in the history of ideas whose traditions have shaped society.

IMAGES OF MATHEMATICS IN PHILOSOPHY

God as a Mathematician

One of the central mathematical innovations of the seventeenth and eighteenth centuries was the development of the infinitesimal calculus, above all advanced by Gottfried Wilhelm Leibniz and Isaac Newton (see Chapter 3). However, both were not only important as working mathematicians, but also as philosophers who developed powerful images of mathematics. Newtonianism is dealt with in Chapter 4, and therefore the focus here will be on Leibniz's reflections on mathematics. For Leibniz, theology and mathematics, metaphysics and logic were closely connected—a connection that also influenced his mathematical practice. For Leibniz, God is not only a rational God, but he is identical with the highest reason, logic. His logical-mathematical, divine mind is both the basis of existence and the basis of knowledge of the world: "When God calculates and exercises his thought, the world is made" (Leibniz [1677] 1989: 185, ann. 4). Since the universe is an incarnation of logic, logical structures are not simply formal entities, but mirror the structures of reality created by God.

According to Leibniz, mathematics is a part of logic—and thus a field of knowledge that man shares with God. Leibniz saw pure mathematics in particular as the application of logic to size, number, measure or weight, thus mediating between the metaphysical and the physical world. He regarded the universe "as a harmonious functional relationship between an infinity of variables" (Loemker 1989: 29f.). This image of mathematics as the key to the universe, which remained effective into the eighteenth century and beyond, is found in probably its most famous formulation in Galileo Galilei:

> Philosophy is written in this grand book—I mean the universe—which stands continually open to our gaze, but it cannot be understood unless one first learns to comprehend the language and interpret the characters in which it is written. It is written in the language of mathematics, and its characters are triangles, circles, and other geometrical figures, without which it is humanly impossible to understand a single word of it; without these, one is wandering about in a dark labyrinth. (Galilei [1623] 1960: 183f.)

Starting from the identity of logic and world structures, Leibniz developed the concept of a symbolic cognition following the example of mathematics. He was convinced that thinking can only take place through the use of signs, such as words or numbers. This thought led to his project of a *characteristica universalis*, an ideal language, which transports the logical connections and the order of ideas and thus maps thought systems onto a system of signs and symbolizes thought processes by reordering the signs. The generation of new knowledge, which he called the *ars inveniendi*, as well as the creation of new technical tools would then take place as a rule-guided combination of signs, similar to calculating in mathematics. On the development of this combinatorial universal language, Leibniz states:

> I should venture to say that nothing more effective can well be conceived for perfecting the human mind and that if this basis for philosophizing is accepted, there will come a time, and it will be soon, when we shall have as certain knowledge of God and the mind as we now have of figures and numbers and when the invention of machines will be no more difficult than the construction of geometric problems. (Leibniz [1675] 1989: 166)

Leibniz's idea of a mathematically inspired art of invention also exerted a great fascination on both mathematicians and philosophers. Christian Wolff, Johann Heinrich Lambert and others pursued the idea further. However, Leibniz's assumption of a universal harmony between autonomous reason and God, the presupposed correlation of mathematical-logical and metaphysical knowledge, became more and more doubtful in the course of the eighteenth century. The background for this was formed by decisive changes in ideas concerning the order of knowledge. Instead of a science subservient to theology, mathematics would increasingly be seen as an autonomous field (see the next section).

This evolution in the image and role of mathematics is also witnessed in the field of mathematics itself. When Leibniz's idea of a *characteristica universalis* was taken up a century later by Carl Friedrich Hindenburg, it was without the overtones of metaphysics. Hindenburg worked on the idea of a combinatorial analysis, that became one of the most influential mathematical fields in Germany during the late eighteenth century. His approach is an important example of the interaction between the image and practice of mathematics. In his work *Novis systematis permutationum* . . . (1781), Hindenburg referred directly to Leibniz as a role model, but his ambition was more delineated. Combinatorial analysis was not intended as a universal grammar of knowledge, but as a general mathematical terminology. His approach provided formulas and algorithms for the arithmetical treatment of power series in order to deal with the concept of functions on the grounds of combinatorial operations. Though informed by the latest developments in analysis, combinatorial analysis was seen by his contemporaries as an alternative to the French-style infinitesimal calculus, announcing the dawn of a genuine German style of mathematics. As such, it was quite enthusiastically received by non-mathematicians too (see the second part of this chapter).

After Hindenburg's death combinatorics as a field of mathematical research would lay fallow for much of the subsequent century and appear a kind of dead end. From a cultural-historical perspective, however, this lack of success is irrelevant: following the sociology of scientific knowledge, cultural history of science should "be impartial with respect to truth and falsity, rationality or irrationality, success or failure" (Bloor 1991: 7). Indeed, also unsuccessful claims of knowledge deserve to be explained and studied, more even, they may help to inform us about the processes at work in vindicating results generally.

Mathematics in the Order of Knowledge

The order of knowledge underwent profound changes during the eighteenth century—and so did the position of mathematics within its hierarchy. One of the most important efforts was the *Encyclopédie ou Dictionnaire raisonné des sciences, des arts et des métiers* (*Encyclopedia, or a Systematic Dictionary of the Sciences, Arts, and Crafts*, 1751–80), an ambitious multi-volume reference work edited by the philosopher Denis Diderot and the mathematician Jean le Rond d'Alembert. Their aim was not a neutral reference work or compendium. Instead, the introduction, d'Alembert's *Discours préliminaire* (*Preliminary Discourse*, 1751), presented a systematic account of human knowledge which executed "a single strategy," namely to ban theology from its dominating role in the intellectual world. In the words of the historian Robert Darnton, "it succeeded in dethroning the ancient queen of the sciences and in elevating philosophy to

Hamburgisches
Magazin,
oder
gesammlete Schriften
zum
Unterricht und Vergnügen
aus der Naturforschung
und
den angenehmen Wissenschaften überhaupt.

Des ersten Bandes erstes Stück.

Hamburg, bey Georg Christian Grund. 1747.

FIGURE 7.1. Title page of the first issue of the *Hamburgisches Magazin*. Photo courtesy of John Carter Brown Library / Internet Archive.

her place" (Darnton 1999: 209). The underlying structure of the *Encyclopedia* accordingly followed a rational order of knowledge, which the German illustrator Christian Friedrich Wilhelm Roth allegorized in the *Tree of Knowledge* engraving for the frontispiece of the 1780 edition. As Darnton points out, "the metaphor of the tree of knowledge [. . .] communicated the idea that knowledge grew into an organic whole, despite the diversity of its branches" (Darnton 1999: 194).

But unlike all other arts and sciences, philosophy was not a branch, but grew directly out of the base of the tree trunk which consisted of understanding and reason, thus emphasizing its most fundamental importance. Perched on the first thick branch originating from philosophy's trunk were the natural and mathematical sciences, including astronomy. Since antiquity, together with arithmetic, geometry, and music, astronomy had been part of the quadrivium of the liberal arts, which were considered a propaedeutic for philosophical knowledge. Within the systematic order of the *Encyclopedia*, this subordinate relationship changed. The mathematical sciences gained a more prominent and independent position. In Roth's illustration, they derived almost as directly from reason as the new rationalist philosophy. And even this level of equality with the leading discipline did not remain unchallenged, especially and not surprisingly, by mathematicians.

Since the early modern period, one of the mathematical sciences, astronomy, had ascended in the order of knowledge as mathematical descriptions of the universe increasingly achieved the explanatory primacy that philosophy and theology once possessed. By the eighteenth century, astronomy was no longer a mere preparatory stage on the path to philosophical understanding. With the help of mathematics, it had become a science that promised answers to the secrets of the universe itself. A telling example is the essay *Das Lob der Sternkunst* (*Praise of Astronomy*), published in 1747 by Abraham Gotthelf Kästner. During his time, Kästner was not only a well-known mathematician—during his lifetime his reputation was considered to equal Leibniz's in Germany—but also the author of literary works in verse and prose which often referred to mathematicians and mathematical themes (cf. Dyck 1980). He stated:

> I do not intend to claim that astronomy is preferable to all other parts of mathematics and scholarship. Only metaphysics has received the title of a queen of sciences, but that was in the barbaric centuries. [. . .] When I have demonstrated the perfection of astronomy, and a hierarchy will be firmly established, it will be easy to assign it to its proper place. (Kästner 1747: 207)

And Kästner leaves no doubt that the appropriate role for astronomy in a new enlightened order of knowledge would be to replace metaphysics as queen of the sciences from "barbaric centuries."

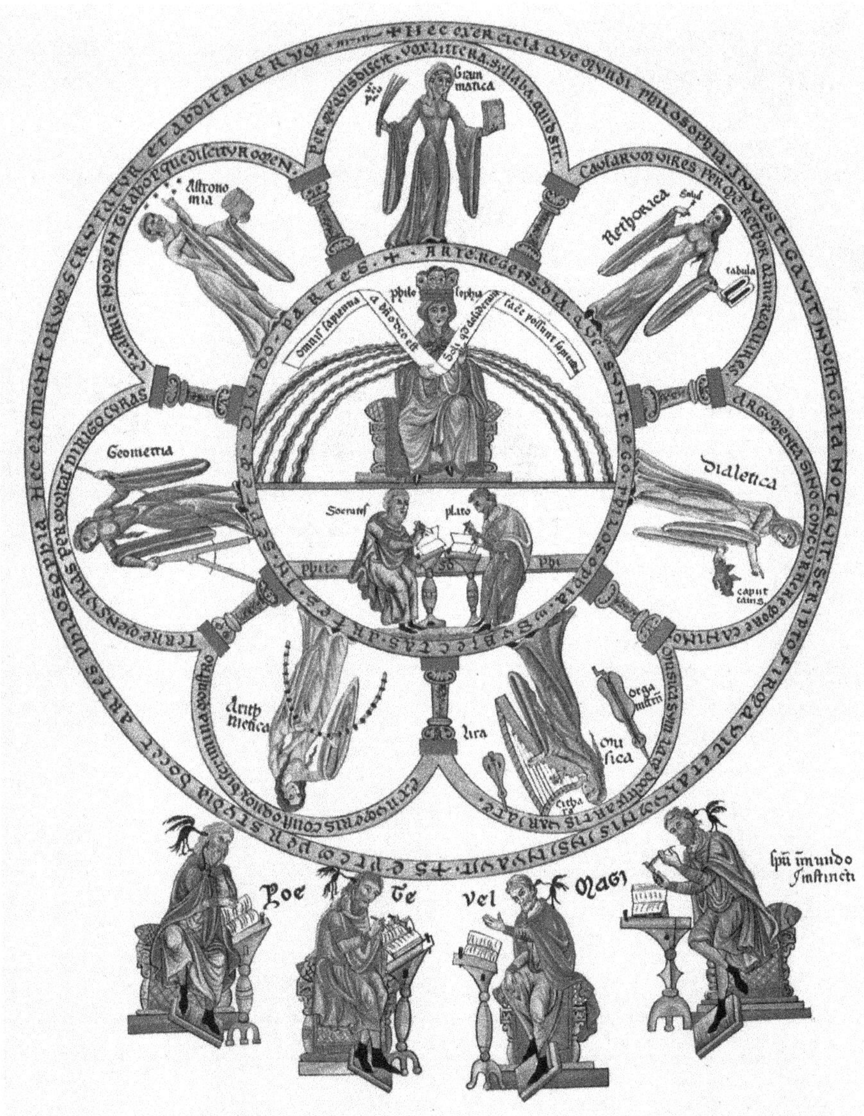

FIGURE 7.2. Philosophy and the seven arts as depicted in the *Hortus Deliciarum* (*c*. 1180). Photo courtesy of Dnalor_01 / Wikimedia Common.

A focal point of the negotiations on the relationship between mathematics and philosophy was the Berlin Academy. Initiated by Gottfried Wilhelm Leibniz, the Berlin Academy was founded in 1700 as the Society of Sciences of the Elector of Brandenburg, later renamed the Royal Prussian Academy of Sciences in Berlin. To this day the Academy—now named for both Berlin and Brandenburg—is an

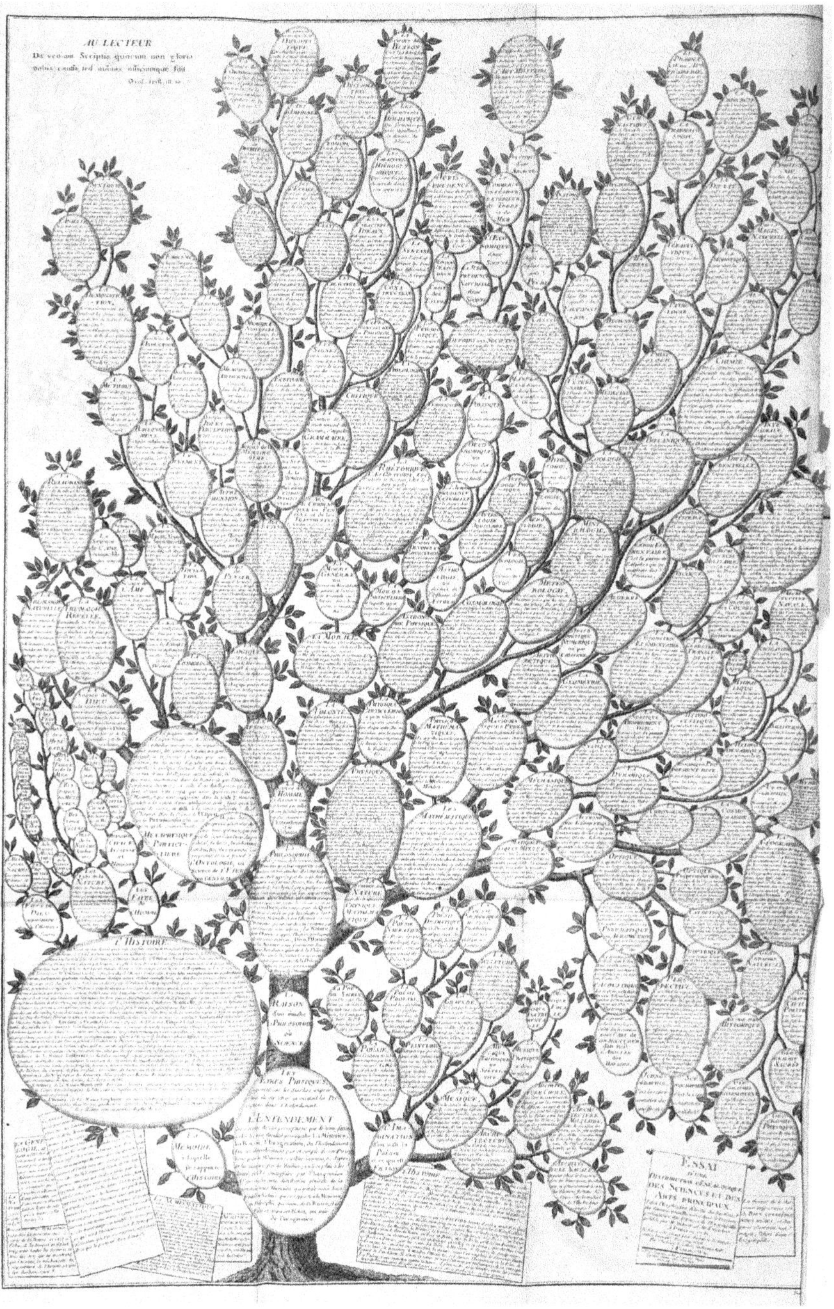

FIGURE 7.3. Fold-out frontispiece of the tree of knowledge, or *Essai d'une distribution généalogique des sciences et des arts principaux*, taken from the index to the *Encyclopedie, ou Dictionnaire Raisonné des Sciences, des Arts et des Métiers* (1780). Photo by SSPL / Getty Images.

important research catalyst for the humanities and the sciences. In the 1740s, the Academy was divided into four classes, which represented the variety of its research areas: experimental philosophy, mathematics, speculative philosophy, and belles lettres. But the boundaries of these fields were not (yet) clearly defined, and their relations towards each other were constantly negotiated within a process of methodological self-assertion: "Interchanges between the classes of experimental philosophy and mathematics were an ongoing challenge to the speculative philosophers" (Laursen 1999: 491).

In 1750, for instance, Euler published his essay *Réflexions sur l'espace et le temps* (*Reflections on Space and Time*) in the proceedings of the speculative class and challenged the metaphysical foundation of mathematics. He argued that the philosophical conceptualization of time and space as products of the human imagination were not suitable notions for the real phenomena that mathematical mechanics dealt with. To him, it was evidently clear:

> that these ideas are not entirely imaginary, as the metaphysicians claim. One should rather conclude that both absolute space and time, as mathematicians represent them, are real things that exist beyond our imagination, since it would be absurd to support the idea that pure imagination could serve as a foundation for real principles of mechanics. (Euler 1750: 2)

Evidence and Certainty of Mathematics

An important ongoing discussion among the members of the Academy dealt with the issue of the evidence and certainty of mathematics (see Chapter 4 in Volume 3). The topic ultimately led to a philosophical debate as to whether it was possible to harness the potential of mathematics in order to further improve philosophical epistemology. In 1761, the theologian and philosopher Johann Georg Sulzer formulated the Academy's annual prize question for 1763:

> Whether metaphysical truths in general, and in particular the first principles of *Theologiae naturalis* and morals, are capable of the same clear proof as geometrical truths, and, if they are not capable of the said proof, then what is the real nature of their certainty, to what sort of degree can one bring their certainty, and whether this degree is sufficient for complete conviction? (transl. Guyer 2000: 17)

The background was formed by the topical attribution of mathematical knowledge as secure and therefore undisputed knowledge. This image of mathematics has a long history, ranging from antiquity to early modern times and Enlightenment, even persisting in contemporary representations of the history of mathematics. It was based on the idea that the so-called mathematical method, as exemplified by Euclidean geometry, produces stable, historically unbound knowledge that is constantly increasing. The mathematician and philosopher Christian Wolff had been a forceful advocate of this approach to philosophy. The philosopher Moses Mendelssohn wrote in his answer to the Academy's prize question, the treatise *Abhandlung über die Evidenz in den metaphysischen Wissenschaften* (*On Evidence in Metaphysical Sciences*, 1764):

> Matters are completely different, however, when it comes to the nature of mathematics. Although greater progress has been made in mathematics than in any other science, the works of the ancients have not, on that account, completely lost their usefulness. Mathematics has its certainty to thank for this prerogative. Its evidence is so great that one could rarely wander far from the truths. One knew less but what one knew were undeniable truths. The discoveries of the moderns have infinitely expanded the boundaries of science, but they have left unaltered the small region that they found already present. (Mendelssohn [1764] 1997: 254)

Mathematics was thus regarded as a science that was free of controversy and on which all members of the scientific community worked harmoniously together to advance their subject. It is worth mentioning that, contrary to this widely held notion, there have of course been controversies in the history of mathematics, such as the dispute between Euler, d'Alembert and Bernoulli over the adequate conceptualization of a mathematical function (see Chapter 3). In the course of the eighteenth century, the image of mathematics as a science with universal validity also acquired a normative function. It no longer served solely to describe the special features of mathematics. Instead, as the Academy's prize question indicates, the mathematical method was also considered to be applicable in other disciplines in order to achieve a comparable objective certainty of knowledge.

This wide-ranging cultural interest in mathematics was sparked by the rise of what was known at the time as New Analysis, a mathematical field that grew out of Descartes' analytic geometry, the aforementioned infinitesimal calculus developed by Leibniz and Newton, and the algebraic basis provided by Euler and Lagrange, to name only a few of its most important stations. These tools, which shape mathematics to this day, formalized concepts and deductive systems that came to have broad applications in the natural sciences. This gave rise to the hope that the methods of the New Analysis might also be fruitful for problem solving in completely different areas. The philosopher and mathematician Nicolas de Condorcet, in his *Esquisse d'un tableau historique des progrès de l'esprit humain* (*Outlines of an Historical View of the Progress of the Human Mind*, 1794), wrote that the "algebraic language [. . .] alone is accurate and truly analytic," and "includes the principles of a universal instrument applicable to all possible combinations of ideas" (Condorcet [1794] 1795: 271f).

While mathematical methods actually played an increasingly important role in the natural sciences, which began to constitute themselves as disciplines in the course of the eighteenth century, the two most famous answers to the Academy's prize question highlighted fundamental differences between mathematics and philosophy and thus demonstrated the methodological limits of mathematics. Moses Mendelssohn, who won the prize, substantiated his argumentation by introducing general comprehensibility as a criterion for the quality of an insight, in addition to certainty:

> Part of the evidence of a truth is, in addition to certainty, perspicuity. To say that a truth is perspicuous is to say that anyone who has ever grasped the proof must immediately be fully convinced of the truth and so set at ease that he does not feel within himself the slightest resistance to assuming it. (Mendelssohn [1764] 1997: 255)

On the one hand, this criterion allowed to restrict the proclaimed general evidence of mathematical theorems to certain areas and to deny the evidence of the more recent New Analysis:

> The first principles of differential calculus are just as undeniable as geometrical truths but they are not as evident or as perspicuous. Hence, the evidence of geometric truths cannot be attributed to them. (Ibid., 255)

On the other hand, by distinguishing between certainty and comprehensibility, Mendelssohn was able to confirm the scientific claim of philosophy. For the metaphysical truths are characterized by the same certainty as the geometric truths, but their complex conclusions are—similar to the differential and integral calculus—not easily understandable for the layman.

The further development of mathematics is reserved for the mathematicians who, according to Mendelssohn, are characterized by impartial minds and direct their research exclusively towards objectively valid inner-mathematical truth. This description also alludes to the stereotype of the unworldly mathematician and the image of mathematics as a non-political science. In explicit contrast to mathematics, Mendelssohn did not regard worldly wisdom as a field of knowledge reserved exclusively for experts. Following an enlightened rejection of ideological despotism, every philosophical layman is instead called upon to join in even if prejudices obstruct his view of rational truth—in this he differs in kind from the mathematician, but only in degree from the trained prejudice:

> Since not everyone has the capacity to examine the philosopher's theses, it is better for such an individual to judge in accordance with his meager insights

> than to recognize and blindly follow some philosophical pope wherever the latter wants to lead him. (Ibid., 278)

Mathematics as Constructions in Pure Intuition

The second prize in the competition announced by the Berlin Academy for 1763 was awarded to Immanuel Kant, probably the most important philosopher of the German Enlightenment, for his study *Untersuchung über die Deutlichkeit der Grundsätze der natürlichen Theologie und Moral*, (*Inquiry Concerning the Distinctness of Natural Theology and Morality*, 1764). Kant distinguished sharply between mathematical and philosophical knowledge: "Mathematics arrives at all its definitions synthetically, whereas philosophy arrives at its definitions analytically" (Kant [1764] 1992: 248). The mathematical method, as Kant explained this distinction further, consists in the active mental construction of new notions as an "*arbitrary combination* of concepts" (Kant [1764] 1992: 248), whereas the philosophical method consists in the explication of already existing notions:

> It is the business of philosophy to analyse concepts which are given in a confused fashion, and to render them complete and determinate. The business of mathematics, however, is that of combining and comparing given concepts of magnitudes, which are clear and certain, with a view to establishing what can be inferred from them. (Ibid., 250)

Kant thus considered an application of the analytical method in the field of mathematics to be just as misguided as an application of the synthetic method in the field of philosophy. In his major work, the *Kritik der reinen Vernunft* (*Critique of Pure Reason*, 1781/1787), however, he abolished this difference between mathematics and philosophy and established a universal form of scientific judgment for all disciplines: according to the Kant of the *Critique*, all scientific knowledge consists of synthetic a priori judgments. This means that they are both objectively true without empirical verification while creating new knowledge not already contained in the original concepts. "The heart of Kant's views on the nature of mathematics is the thesis that the judgments of pure mathematics are synthetic a priori" (Kitcher 1992: 109). Mathematics thus provides an example that such forms of judgment are not only theoretically conceivable, but actually exist: "Mathematics gives the most resplendent example of pure reason happily expanding itself without assistance from experience" (Kant [1781/1787] 1998: 630). The objective truth of mathematical judgments is thereby guaranteed by the cognitive faculty of pure intuition, the transcendental condition of sensibility. However, these means are not available to metaphysics, which Kant assigned to the discursively operating cognitive faculty of pure reason: "*Philosophical* cognition is *rational cognition* from *concepts*, mathematical cognition that from the *construction* of concepts. But to *construct* a concept means to exhibit *a priori* the intuition corresponding to it" (Kant [1781/1787] 1998: 630). Consequently, mathematical reasoning cannot be successfully employed by philosophy—as well as the other way round. Kant asserted:

> Mathematics is thoroughly grounded on definitions, axioms, and demonstrations. I will content myself with showing that none of these elements, in the sense which the mathematician takes them, can be achieved or imitated by philosophy; and that by means of his method the mathematician can build nothing in philosophy except houses of cards, while by means of his method, the philosopher can produce nothing in mathematics than idle chatter, while philosophy consists precisely in knowing its bounds, and even the mathematician, if his talent is not already bounded by nature and limited to his specialty, can neither reject its warnings nor disregard them. (Ibid., 637)

In contrast to Christian Wolff, Johann Heinrich Lambert and other philosophers of the eighteenth century, Kant bid farewell to the utopia of a mathematization of metaphysics. Instead, he conceived philosophy and mathematics as separate sciences with specific but disjoint cognitive interests and means.

Kant's description of mathematical reasoning is modeled on the aforementioned method of proof applied in Euclid's textbook on geometry, which served Kant as a prototype of the mathematical method of knowledge. The increasingly algebraicized mathematical practice of the time contradicted this image strongly. Nevertheless, by connecting mathematics to pure intuition instead of identifying it with the discursive faculty of reason, Kant drew a decisive new image of mathematics. Whereas at the beginning of the eighteenth century, Leibniz had described mathematics as divine knowledge, Kant now presented it as constructions of the human mind. The image of mathematics was thus by no means fixed. Contrary positions resurfaced once again, for example, at the beginning of the twentieth century during the so-called fundamental crisis of mathematics, whose philosophical thread discussed whether mathematical knowledge is discovered or invented.

Mathematics and Imagination

The image of mathematics presented by Kant is also an important stage in the history of imagination, which includes the ever-changing relationship between mathematics and the realm of aesthetics. Conceived as an integrative mental force overcoming the dualism of sensation and reason, imagination played an important role in how the relationship between science, art and poetry was modeled in the eighteenth century.

The role of imagination in eighteenth-century thought has long been stereotypically represented, imagination being responsible for a radical separation of the realms of art and science. According to this view, the sciences relied above all on rational reasoning and encountered imagination with "Fear and Loathing" (Daston 1998). In contrast, art had banned reason and devoted itself to an exaggerated, quasi-mystical power of imagination. But in fact, for the eighteenth century, imagination was as important for Isaac Newton as it was for Alexander Pope. Imagination has been conceived of as an ambivalent mental faculty since antiquity. While Aristotle does not ascribe any direct epistemic power to *phantasia*, he considers it to be an indispensable component of the acts of thought realized by reason—not least in mathematics. In the early modern times, Descartes and others subordinated imagination to reason in order to prevent its abuse, not to banish it completely from the process of knowledge, in which it instead has a productive function when used correctly. Accordingly, Descartes promoted a balanced interplay of imagination and reason, whereby the power of imagination could be safely practiced above all in mathematics. This conception remained valid through the eighteenth century. Kant defines productive imagination as "the faculty for representing an object even without its presence in intuition," which mediates "between the sensibility and the understanding by synthesizing sensible representations in accordance with the categories" (Hall 2011: 220). It is thus not only fundamental to mathematical reasoning but also to the representation of aesthetic ideas. D'Alembert followed a similar line of thought in his *Preliminary Discourse*: For him, imagination is closely related to the faculty of abstraction. Therefore, it is not only of the highest importance for the fine arts and poetry, but also for the sciences, and to an even greater extent, the more abstract their reasoning is. A paradigmatic example is provided by mathematics, in a way similar to Kant, and especially by geometry, in which "imagination plays the greatest part." D'Alembert states polemically:

> I ask pardon of those superior wits who are detractors of Geometry; doubtless they do not think themselves so close to it, although all that separates them perhaps is Metaphysics. Imagination acts no less in a geometer who creates than in a poet who invents. [. . .] Of all the great men of antiquity, Archimedes is perhaps the one who most deserves to be placed beside Homer. (D'Alembert [1751] 2009: 61–3)

Apart from geometry, it is above all astronomy in which not only imagination and reason, but also pleasure and moral sentiment come together in an ideal way. In his *Abrégé d'Astronomie* (*Compendium on Astronomy*, 1774), Joseph Jérome de la Lande describes its advantages:

> The wonders discovered there captivate the soul and occupy it in a noble, delightful manner that is free from danger; they elevate the imagination; they perfect the mind; they fill and satisfy the heart; they ward off dangerous and frivolous desires; they constantly provide new pleasure. (Lalande 1774: x)

This fascination with astronomy in particular, with the sensual-aesthetic qualities of mathematics and its epistemic potential in general was shared by some poets in the eighteenth and early nineteenth centuries, whom d'Alembert wrongly accused of having a low regard for mathematics. This applies especially and surprisingly to Early German Romanticism at the end of the eighteenth century, which has long been wrongly described as an anti-scientific and above all irrational movement. Instead, it is unthinkable without the Enlightenment, whose rationalist tendencies it takes up both critically and constructively. Kant's attribution of mathematics to the cognitive faculty of intuition is particularly influential here, since the poets make mathematics their subject primarily as an aesthetic form of knowledge.

IMAGES OF MATHEMATICS IN LITERATURE

A Sensuous Logic

Friedrich Schlegel, the leading mind of early German Romanticism, valued mathematics for the aesthetics of its formal language and deployed it to make his philosophical propositions capable of being experienced on a sensory level. Schlegel frequently made use of mathematical notation to express philosophical ideas. For instance, in his posthumous *Fragmente zur Litteratur und Poesie* (*Fragments on Literature and Poetry*), one can find passages like these (Schlegel [n.d.] 1981):

$$\frac{F}{0}\frac{S}{0}\frac{M}{0}\ \textit{are}\ \text{the poetic ideas.} - \langle \textit{The}\ \pi\ \textit{ideal} = \sqrt[\frac{1}{0}]{\frac{FSM\frac{1}{0}}{0}} = \textit{God}\rangle.$$

These notations may seem cryptic, but they follow a system. The individual Latin and Greek letters are abbreviations for philosophical concepts. In this formula, F stands for "fantasy"; S, for "sentimentality"; M, for "mimesis"; and π, for "poetic." The division by zero expresses the absolute, or infinite, form of the numerator. The exponent and square root functions represent the move from unity to manifold and from manifold to essence, respectively. Accordingly, Schlegel's formula can be translated as follows:

$\frac{F}{0}\frac{S}{0}\frac{M}{0}$ *[absolute fantasy, absolute sentimentality, absolute mimesis] are the* poetic ideas. — ‹*The* π *[poetic] ideal* $= \sqrt[\frac{1}{0}]{\frac{FSM\frac{1}{0}}{0}}$ *[the absolute essence of the absolute poetic ideas raised to the manifold]* = *God*›.

In his *Vorlesungen über Enzyklopädie* (*Lectures on Encyclopedia*, 1803), August Wilhelm Schlegel, brother of Friedrich Schlegel and himself a poet and literary critic, distinguished strictly between philosophical and mathematical communication. The first he considered an art, the latter only a mechanism. To him, mathematics was less a science than a useful craft consisting solely of aids and shortcuts (Schlegel [1803] 2006: 21–3). If his brother Friedrich Schlegel had agreed with him, in the best case, his notation would merely be a system of shorthand terms and memory aids. In the worst case, it would be a generator of the ubiquitous use of pseudo-mathematical magic formulas that for instance Johann Wolfgang Goethe criticized as bereft of epistemic value. Friedrich Schlegel saw it differently, however. The notation, he maintained, is a way of doing philosophy that is superior to syllogistic and analytical reasoning. In contrast to the "emptiness and tautology of logical principles," his mathematically inspired formulas aim to advance knowledge, because they represent "objects by their inner composition, their elements, their gradual development, and their internal relationships" (Schlegel [1804/05] 1964: 322f.). Schlegel understands mathematics as a "sensuous logic" that "relates to philosophy as the material arts, music, and sculpture relate to poetry" (Schlegel [1798] 1967: 232). His notation is, therefore, a "*method of construction*." It is also "temporary and hypothetical," "valid solely for subordinate, technical applications." Schlegel aspires to an ideal philosophy that has yet to come, a philosophy in which all forms and methods will be adequate to their content, a thinking that accords with both its object and its representation, much like poetry itself (Schlegel [1804/05] 1964: 323, 386, original italics). Mathematical formulas, Schlegel believed, would serve as the stepping stone to this ideal form.

Michel Chaouli has argued that Schlegel's vision of an ideal philosophy draws even more on chemistry than it does on mathematics (Chaouli 2002). Efforts to understand and classify the chemical elements in the late eighteenth century brought combinatory logic to bear on the idea of elective attraction, a theory developed by contemporaneous scientists to explain why certain elements attract each other while other combinations do not occur. Schlegel associated the vagaries of observed chemical attractions with empirical contingency. He believed that the job of philosophy, which he once termed "logical chemistry" (Schlegel [1796–8] 1967: 33), was to understand the interactions between the empirically particular and the rationally abstract. The language of philosophy consists of "*symbol[s]* for *intellectual intuition*" that are expressible neither as concepts nor as representations of their construction. Each of its symbols "contains something inexplicable, irresolvable, incomprehensible, not in the sense that it cannot be absorbed by human consciousness" but, rather, in the sense "that it necessitates intellectual intuition" (Schlegel [1804/05] 1964: 389, original italics).

Therefore, the unique aesthetic qualities of Schlegel's formulas are to be taken seriously as part of the process of finding an adequate philosophical language. They can indeed be translated into natural language, but natural language cannot capture the spatial arrangement of their symbols. There remains an unbridgeable gap between what can be said and what can be seen. Natural language and diagrammatic formulas are different modes of representation and different modes of knowledge. Trying to think about them concurrently triggers a moment of consternation, a feeling akin to something "inexplicable, irresolvable, incomprehensible," something that is accessible neither by grammar nor by logic alone. In this regard, the epistemic value of Schlegel's formulas is that they allow us to get a fleeting glimpse of what unrepresentability might mean. Accordingly, the function of Schlegel's formulas resembles that of the philosophical or literary fragment in German Romanticism: a means of temporarily exceeding the limits of thought.

An Ingenious Art of Invention

Friedrich Schlegel corresponded closely with the early Romantic Friedrich von Hardenberg, better known by his nom du plume, Novalis. In addition to being a philosopher and poet, Novalis had extensive training in mathematics, which played a central role in his thought and writing, especially in his encyclopedia project. Inspired in part by Diderot's and d'Alembert's aforementioned *Encyclopedia*, Novalis had sought to assemble his copious musings on art, religion, and science into a single encyclopedic work. He never completed it, but a draft was later published as *Das Allgemeine Brouillon* (*Notes for a Romantic Encyclopedia*, 1798/99). The work collects some 1200 numbered entries that gloss theories and concepts from an enormous range of disciplines. It is here that Novalis' most important thoughts on mathematics can be found. For example, entry 1126 reads:

> Mathematics is genuine science—because it contains *created knowledge*—the products of its own spiritual activity—and because it *methodically inspires*.
>
> It is art, because it has fashioned inspired procedures into rules—because it *teaches* one to be a genius—and because it replaces *Nature with reason*. (Novalis [1798–99] 2007: 186)

Novalis here characterizes mathematics in two ways. First, he sees in it the active generation of new knowledge. This idea of mathematics follows Immanuel Kant, whose theories Novalis takes up at numerous points in his writing. The second part of Novalis' characterization holds that mathematics is an art. This understanding draws on the inclusion of arithmetic and geometry in the seven liberal arts of classical antiquity and the status of mathematics in the eighteenth century as an ancillary science that was the product of strict, learnable rules. According to Novalis, then, the mathematician is both: a divine creator of nature and a craftsman who forges objects. The first aspect ties in with the eighteenth-century notion of the poet as creative genius, who generates new ideas that do not derive from any rule, predicates traditionally reserved for God. For Kant, genius "consists in the happy relation, which no science can teach and no diligence learn, of finding ideas for a given concept on the one hand and on the other hitting upon the *expression* for these" (Kant [1790] 2000: 194). "The proper field for genius," Kant writes, "is that of the power of imagination, because this is creative and, being less under the constraint of rules than other faculties, it is thus all the more capable of originality" (Kant [1798] 2007: 329). Kant is emphatic in excluding scientists—mathematicians in particular—from the realm of genius:

> Thus everything that Newton expounded in his immortal work on the principles of natural philosophy, no matter how great a mind it took to discover it, can still be learned; but one cannot learn to write inspired poetry [. . .]. The reason is that Newton could make all the steps that he had to take, from the first elements of geometry to his great and profound discoveries, entirely intuitive not only to himself but also to everyone else. [. . .] In the scientific sphere, therefore, the greatest discoverer differs only in degree from the most hard working imitator and apprentice, whereas he differs in kind from someone who is gifted by nature for beautiful art. (Ibid., 187f.)

By contrast, Novalis takes the synthetic *a priori* character of mathematical positions to demonstrate their originality and genius. In doing so, he places mathematics in the realm of creative invention:

> Everything *Real* created out of *Nothing* (like numbers and abstract expressions for instance)—has a wonderful affinity with the things of another world—with the infinite series of singular combinations and relations—with a mathematical and abstract world in itself as it were—with a *poetical mathematical* and abstract world. (Novalis [1798–99] 2007: 160)

For Novalis, mathematics is an inventive art that arises from both genius and method, and he believes that its eighteenth-century advances promoted new knowledge in other areas. This belief is predicated on the assumption of a fundamental analogy between all the sciences. In his *Notes for a Romantic Encyclopaedia*, Novalis expands on Kant's idea of the synthetic *a priori*:

> Kant's question: "Are *a priori* synthetic judgments possible?" may be specifically expressed in the most varied manner.
>
> E.g. = Is philosophy an *art* (dogmatics) (science)?
> = Is there an art of invention devoid of data, an absolute art of invention?
> = Can diseases, etc. be created at will?
> = Can we conceive verse according to rules, and an insanity according to fundamental principles?
> = Is *perpetual motion*, etc. possible?
> = Is genius possible—can genius be defined?
> = Can the circle be squared?—
> = Is *magic* possible?
> = Can God, freedom, and immortality be demonstrated?
> = Does a calculation of the infinite exist?
>
> and so on. (Ibid.: [1798–99] 2007: 119)

The equal signs indicate that, for Novalis, his reformulations are synonymous with Kant's original question, or at least sufficiently similar to justify comparison. Each is supposed to describe the same core problem from the vantage point of a different discipline—philosophy, medicine, poetry, psychology, physics, and mathematics. There is, however, an important difference between the questions concerning mathematics and the questions concerning the other subjects. This is because, according to Novalis, the mathematical questions—Can the circle be squared? Does a calculation of the infinite exist?—could already be answered affirmatively in his day. The calculation of the infinite became possible with the development of calculus in the early eighteenth century. As for the first question, calculus could not square the circle in the sense intended by ancient geometers. But, Novalis reasoned, the techniques of infinitesimal calculus, which determines areas under curves by approximating the infinite sum of ever smaller rectangles, might come close:

> The basis of this is the hypothesis or postulate that the square is the archetype of the *circular form*. The problem of the circle is therefore the problem of reducing all figures to the square—or *inversely*, of all figures to something round. The greater the number of divisions we make of this figure—the more precise the solution obtained. An infinite number of divisions gives us an infinitely precise solution. Differential and integral calculus. (Novalis [1798–99] 2007: 73)

Calculating Celestial Beauty

The privileged place of mathematics as an archetype of the *ars inveniendi* also finds expression in Novalis' fiction, particularly in Klingsohr's fairy tale. The figure of Klingsohr appears in Novalis' unfinished novel *Heinrich von Ofterdingen* (*Henry of Ofterdingen*, [1802] 1842). At the end of the first part of the novel, Klingsohr, a celebrated poet and teacher, relates a fairy tale that he wrote about the rise of a golden age. Among the tale's colorful allusions to a variety of myths from around the world are references to contemporaneous recent discoveries in the natural sciences such as galvanism, magnetism, and chemical bonds. Such is the prominence of these themes that scholars consider Klingsohr's fairy tale to be an early form of science fiction. At the same time, Novalis seemed to suggest that abstract mathematics is the foil of poetry, since it is only after the downfall of mathematics that the golden age finally begins. The opposition between mathematics and poetry is represented by the figures of the scribe and the child Fable. In an allegorical interpretation of the fairy tale, the scribe embodies reason, and Fable impersonates poetry. Forever resentful and dissatisfied, the scribe seeks to undermine Fable's efforts to emancipate the fairy tale world from its wrongfully inverted order. The antagonism between the figures becomes tangible in their very first scene together. An elderly woman named Sophia—an allegory of wisdom—assays the truth of the scribe's words by dipping his parchment into a bowl of clear water:

> When she dipped the leaves in the water, and found on withdrawing them, that some of the writing remained still glittering, she gave them to the scribe, who fastened them in a great book, and seemed much out of humor when his labor had been in vain, and all the writing had been obliterated. (Novalis [1802] 1842: 159)

Then Fable takes the pen and begins to write. The scribe hands the leaves over to Sophia in the hope that the water might clean them. He is extremely annoyed, however, when he has to realize that not a single one of Fable's words is erased. Fable's ability even goes beyond creating text whose words remain eternal: When Sophia dips her finger in the miraculous water and sprinkles drops of it on Fable, they dissolve

> into a blue vapor, exhibiting a thousand strange images, and floating and changing constantly around them. If one of these by chance touched the scribe, many figures and geometrical diagrams fell down, which he strung with much diligence upon a thread, and hung them for an ornament around his meagre neck. (Ibid.: 159)

It is obvious that the portrayal of mathematics in the fairy tale is the opposite of positive. It represents an ossified understanding whose truths are eternal only in part. It is an understanding that can do nothing with the strange images of poetry other than to analyze them abstractly and thereby deprive them of life. But this interpretation overlooks the many positive remarks that Novalis made about mathematics. For him as for Schlegel, mathematics cannot be reduced to dry abstraction. Rather, it is also part of pure intuition, where it unifies reason and the imagination.

In the opening scene of Klingsohr's fairy tale, we find a poetic image of this inventive harmony. A star-king named Arcturus plays an astronomical card game with his daughter:

> The attendants brought a table and a little casket, containing a heap of leaves, upon which were inscribed mystic figures of deep significance, constructed of constellations. The king reverently kissed these leaves, mixed them carefully together, and handed some to his daughter; the rest he kept. The princess placed them in a row upon the table; then the king closely examined his own, and chose with much reflection before he added one to them. At times he seemed forced to choose this or that leaf. But often his joy was evident, when he could complete by a lucky leaf a beautiful harmony of signs and figures. (Ibid.: 157)

The figures in Arcturus' court are not the lifeless ornaments around the skinny neck of a resentful grouch. As the cards are placed in orderly rows, the constellations they portray join to become larger configurations of stars whose beautiful harmony sparks joy in the players. The king's reflection before adding a card indicates that the beautiful harmony is not only a visual pleasure but also an intellectual achievement. As Arcturus and his daughter continue to play

> a gentle but deeply moving music was heard in the air, seeming to arise from the stars gliding past each other in a wondrous motion, and from the other movements so peculiar. The stars floated round, now slowly, now quickly, in continually changing lines, and curiously imitated, to the swell of the music, the figures on the leaves. The music changed incessantly with the images upon the table [. . .]. With incredible adroitness the stars flew together according to the images. (Ibid.: 157f.)

The music seems to conjure the representations on the cards into actual stars, eliminating the boundary between abstract signs and concrete objects. This scene prefigures the ending of the fairy tale. By the time the golden age has come, Fable and the scribe have vanished from the earth, along with the antagonism between them. Fable ascends to the celestial plane, where, like Arcturus, she creates celestial spectacles. She has become a practitioner of the perfect art of invention that relies on understanding and intuition in equal measure. The fate of the scribe—he is bitten to death by tarantulas—is decidedly less glorious. Nevertheless, he does attain a kind of immortality. The altar that holds the miraculous bowl in which Sophia dipped his writings is eventually remodeled to include a special kind of ornament: a "great circle" on the floor, laid with "variegated stones" and "graced with noble and significant figures" (Novalis [1802] 1842: 182f.). Both an art work and a geometric figure, this mosaic can be seen to serve as an improved version of the ornament that the scribe wore around his neck.

In the happy ending typical of the fairy tale genre, Novalis unifies the mortifying, purely abstract calculation with a sensuous heavenly harmony of the spheres. This juxtaposition points to a change in the image of calculating during the transition from the eighteenth to the nineteenth century. This change is caused by an increase in mathematical education in general as well as by the technical progress of calculating machines and growing use of algorithms: In the course of these developments, calculating lost its ingenious aura, whereas in the eighteenth century it was largely regarded as the embodiment of intelligence: "Prodigious mental reckoning was a topos in the eulogies of great mathematicians, Gauss's lightning arithmetic being perhaps the last of these stock legends" (Daston 1994: 186).

Mathematics versus Poetry

In contrast to Novalis' harmonious incorporation of mathematics into an idealistic utopia stands the critical view of Goethe in his later years, who was especially skeptical about the universal use of mathematics in the natural sciences. In his notes *Über Mathematik und deren Mißbrauch* . . . (On Mathematics and Its Misuse, 1826), he warned natural scientists against a careless *rage du calcul*. The "magic formulas of mathematics," he complained, are often used dishonestly to complicate intuitive phenomena and prevent amateur enthusiasts from obtaining knowledge about nature (Cf. Goethe [1826] 1989: 69). This is the reason that scholars have often portrayed Goethe as completely hostile to mathematics. This characterization is nevertheless false, and Goethe himself rejected it. But it is also true that, unlike the empirical sciences, abstract mathematics was foreign to him. And Goethe was considerably less optimistic about the increasing mathematization of nature than many of his contemporaries, Schlegel and Novalis included. In his novel *Wilhelm Meisters Lehrjahre oder die Entsagenden* (*Wilhelm Meister's Journeyman Years or the Renunciants*, 1821/29), Goethe explored the ethical implications of the mathematical view of nature by way of an astronomical poetry that would feel right at home in Arcturus' court. *Wilhelm Meister's Journeyman Years* first appeared in 1821. The second, final edition was published in 1829. It contains a complex fabric of interlaced narrative strands and genres, presented from multiple perspectives that do not add up to a harmonious whole without considerable interpretive effort. In terms of content, the *Journeyman Years* is a social novel, whose figures seek to realize a better society by means of economic, technological, and industrial innovations.

Though numerous reflections on natural science find expression in the *Journeyman Years*, Goethe omits all discussion of mathematics—explicitly, at least. In chapter 10 of the first book, the protagonist Wilhelm meets the elderly lady Makarie and her companion, an astronomer. The three have come together to talk about mathematics. After some rather extensive preliminaries in which they assure each other of their interest in the subject, the astronomer delivers an introductory lecture. The lecture does not appear in the novel, however. This, the narrator explains, is so as not to bore the reader:

> If, however, we find ourselves disinclined to let the worthy man read, our patrons will most likely be pleased, since what was said earlier against Wilhelm's presence at the conversation applies even more to the situation in which we find ourselves. Our friends have taken a novel into their hands, and if it has already been more didactic here and there than it should be, we find it advisable not to put the patience of our well-wishers further to the test. We intend to have the papers at our disposal printed elsewhere, and shall proceed with our history without further ado [. . .]. (Goethe [1829] 1995: 176f.)

Goethe scholars are in agreement that the omitted lecture consisted of passages from the posthumously published text *On Mathematics and Its Misuse*. Aphorisms based on the lecture are included in materials from the archive that Makarie is supposed to have maintained and that Goethe appended to the novel's end. But Goethe forgoes putting a complete copy of his critical essay on mathematics in the *Journeyman Years*. In this regard, Goethe, unlike the Early Romantics, excluded mathematics from the realm of aesthetics and literature.

Mathematics' Moral Indifference

But mathematics remains present in Goethe's novel as a problem. The form it takes is not discursive but literary. It starts with Makarie, who, it turns out, has a special relationship with the cosmis:

> Makarie stands in a relationship to our solar system that one hardly dares to express. Not only does she harbor it, and see it in her mind, in her soul, in her imagination; she constitutes a part of it, as it were. She sees herself drawn

> along in those heavenly circles, but in her own particular way; since childhood she has moved around the sun, and, to be specific, as has now become clear, in a spiral course, moving ever farther from the center and circling toward the outer regions. (Goethe [1829] 1995: 409f.)

For a long time, Makarie has regarded her feelings as a mental illness. She dismissed them as delusions because she could not reconcile her inner visions with her outer observations of the heavens. But Makarie finally comes to terms with her visions by making sense of them through mathematics. She manages to do this with the help of her companion who combines a characteristic set of different abilities and qualities: he is a doctor, a mathematician, an astronomer, and above all a highly noble person. Only in this combination of abstract knowledge and ethical quality, Goethe suggests, can the mathematician make a valuable contribution to knowledge of the world and approximate the ideal of the perfect human being. Mathematics alone, as Goethe notes in his *Maximen und Reflexionen* (*Maxims and Reflections*, 1833), achieves nothing moral at all. Like rhetoric, it is a craft that is morally indifferent to its subject. The mathematician must compensate for this deficit in his character:

> However, here it depends on the nature of the person carrying out an occupation or an art of this kind. A convincing lawyer in a justified case, a powerful mathematician under the starry heavens—both appear godlike. [. . .]
> The mathematician is only perfect to the extent that he is a perfect human being, that he can inwardly experience the beautiful in the true. (Goethe [n.d.] 1993: 153, trans. David Wood)

Though the astronomer in the novel is primarily driven by scientific curiosity, his character allows him to win the confidence of Makarie, who readily shares her secret visions with him. His is an alternative to the forced mathematical understanding of nature that Goethe criticizes in Newton's theory of colors, whose "ascendancy and consideration has hitherto impeded a free inquiry into the phenomena of colours." This theory does not manage to "guard against the possible abuse of this abstract view, in order that the practical deductions we look to should be really useful" (Goethe [1810] 1840: xxf.). By contrast, the astronomer's calculations do not call into question Makarie's visions; rather, they reveal the truth that dwells in them and ensure that Makarie no longer sees them as an illness. Makarie can enjoy her intuition of the universe while deriving an ethical mission from her relationship to the cosmos. It seems that the astronomer also undergoes a transformation of sorts. His first response to Makarie's visions is in keeping with his profession and cast of mind: "He is a mathematician and therefore obstinate, an enlightened spirit, and therefore skeptical." Resisting Makarie's intuitive connection with the universe, the astronomer seeks a rational explanation, assuming that Makarie's power of imagination simply reproduces astronomical knowledge using "a secret element of calculation" (Goethe [1829] 1995: 411). But he must reject this hypothesis because Makarie's visions anticipate planets only later discovered by astronomy. Perplexed, the astronomer examines Makarie by means of mathematics:

> The scientist then had her describe minutely what she saw, which was only occasionally very clear. He made calculations and concluded from them that she not only carried the entire solar system within her, but also that she moved within it spiritually as an integral part. He proceeded according to this hypothesis, and his calculations were confirmed in an incredible manner by her statements. (Goethe [1829] 1995: 182f.)

Both mathematical and mystical understandings of nature converge in this passage. The convergence is not a utopian harmony in the manner envisioned by Novalis, however. Noticeable fractures frustrate such a possibility. Goethe depicts Makarie as a representative of an intuitive unity of human kind and nature and affords her a special place that is superior to a mere mathematical understanding of the world. For example, it is Makarie's assertions not the astronomer's calculations that ultimately confirm his calculus. But unlike Wilhelm, who intuitively guesses Makarie's unity with the universe in a dream, regards it as a great miracle and is filled with awe, the astronomer ultimately falls back on a mechanistic comparison. From his examination he concludes that Makarie is "a living armillary sphere, a spiritual clockwork, that might be capable [. . .] of following the movement of the stars in its own way" (Goethe [1829] 1995: 411).

On the one hand, the descriptions of Makarie's worldview evoke an experience of the sublime disenchanted by the comedy of being reduced to mere machinery. On the other hand, Goethe puts the professional myopia of the astronomer in an ironic light. But the point of the *Journeyman Years* is not to dismiss mathematical advancement: Referring to optic magnification tools, Wilhelm states: "We can as little ban these glasses from the world as we can ban machinery" (Goethe [1829] 1995: 179). In contrast to this certainty of the future is the sense that Makarie's visions belong to the past. What the reader learns about her mostly comes from fictional archival materials: "Unfortunately, this essay was written from memory, long after its contents were communicated, and thus cannot be regarded as entirely authentic, as might be wished in so remarkable a case" (Goethe [1829] 1995: 409). Goethe's Makarie draws on premodern models of nature to remind us of what was lost once we achieved dominion over nature. But the melancholy this causes is also an appeal to bring

FIGURE 7.4. Armillary Sphere (1731). Photo by SSPL / Getty Images.

ethics backs to technology and mathematical modernity. This hope is suggested by Makarie's visions as they extend beyond the known universe:

> No one's imagination can follow her that far, but we hope that such an entelechy will not withdraw entirely from our solar system, but when she reaches its boundary will long again to return, in order to exercise her influence upon earthly life again, to the benefit of our descendants. (Goethe [1829] 1995: 411)

CONCLUSION

Probably one of the most stable images of mathematics encompasses its certainty and its historical independence: mathematical knowledge—especially in contrast to the humanities—does not seem to be afflicted by political and historical conditions, its truths are not susceptible to ideologies. Undoubtedly, this image has a factual basis in the epistemic specificity of mathematical knowledge and in the observation that mathematics has produced, expanded, and enlarged relatively secure knowledge for centuries.

However, this historical immunity does not apply to the images of mathematics, which are decidedly historically variable and ideologically conditioned. They are important for a cultural history of mathematics on at least three levels of analysis: "(1) a properly disciplinary level; (2) a social and institutional level; (3) finally, a more diffuse cultural level which concerns the place of mathematics in the general configuration of knowledge, the prestige it enjoys and the values it embodies" (Bottazzini and Dalmedico 2001: 4). These levels interlock. A society can appreciate the high degree of mathematical abstraction as the epitome of divine wisdom or as an expression of a rational faculty that is equally suited to all people and should therefore be enhanced. It can be criticized as an elitist gesture of an arcane guild intent on exclusivity or even defamed anti-Semitically as a despicable expression of the "Jewish spirit." Images have found and continue to find counterparts, not least in the institutional organization and promotion of mathematical research. And they render supposedly immune mathematics vulnerable to ideological instrumentalization. Their historically informed analysis is therefore important for a critical perspective on the cultural history of mathematics which might also contribute to a sensitive assessment of the present images of mathematics and their implications.

The poetic reflections on mathematics of Goethe, as well as those of Schlegel and Novalis, illuminate the aesthetic influence of mathematics in a certain era. The eighteenth century marked a productive highpoint in culture and the arts. Authors of canonical works of this century were later to serve as arbiters for a wide variety of worldviews, not least for the mathematical understanding of the world. It may seem far-fetched to claim that an inversely proportional relationship existed between the interest in Goethe's work and the prestige of the exact sciences in German cultural history since the nineteenth century. But the argument of the historian of science Ivo Schneider that Goethe served as an alibi for the low estimation of mathematics among the educated bourgeoisie is significant not only for Goethe scholars but also for those interested in the relations between scientific knowledge and ideologies (cf. Schneider 2011: 252). The same holds true for Early Romanticism which was corrupted in order to serve radical anti-rationalism in the twentieth century—a strategy that can be identified by a historically informed view of the image of mathematics in the eighteenth century.

NOTES

Chapter 3

1. According to Katz (1995: 165–9), the formulas for the sums of the kth powers of integers, at least through $k = 4$, were developed in Egypt some 650 years before Newton by ibn al-Haytham (965–1039), known in Europe as Alhazen.
2. This does not mean that the success of Euler's treatises was immediate. After six years, 406 out of 500 copies of Euler's *Institutiones calculi differentialis* remained unsold (Calinger 2016: 396). It clearly shows the difficulties to find a readership for mathematical books of that level.
3. The Riccati equation $x^m\, d^2x = d^2y + dy^2$, where x and y depend on a further variable p.
4. Namely, $\sigma \frac{\partial^2 y}{\partial t^2} = T \frac{\partial^2 y}{\partial x^2}$, where σ represents the mass per unit of length and T the tension of the string.
5. Find the curve having a given length and enclosing the maximum area.
6. Recall that the quintic cannot be solved algebraically.

Chapter 4

1. The first part of this chapter was written by Jip van Besouw.
2. The second part of this chapter was written by David Bellhouse.
3. It would be a mistake to conclude that it was games of chance alone, or gambling in general, that initially motivated mathematicians to study probability. Likewise, it would be a mistake to conclude that gambling played no part in the early development of probability theory. This can be illustrated by an analogy to the music world. By the late 1750s, when games of chance were fading from research into probability theory, they were making inroads into music composition. In 1757, the composer Johann Philipp Kirnberger constructed a musical dice game in which bars of music

were chosen by the throw of the dice and then put together to make the entire composition. Later, other composers, including Wolfgang Amadeus Mozart, wrote these kinds of pieces. The ubiquity of dice games in the eighteenth century was a motivating factor in the emergence of this kind of music. At the same time, it was a challenge to the composer, perhaps a minor one to a talented composer, to write fragments of music that would fit well together in the completed piece.

Chapter 6

1. All references to Euler's papers are given as Exxx, with xxx being the number of the paper in the Eneström index that lists all of Euler's publications. These papers can be found in the online Euler Archive.

Chapter 7

1. Since the designations *body* and *image* seem to be more intuitive and more easily distinguishable from each other, they are here given preference over Carson's *interpretation* and *implication*.

BIBLIOGRAPHY

Albury, William R. (1978), "Halley's Ode on the *Principia* of Newton and the Epicurean Revival in England," *Journal of the History of Ideas*, 39 (1): 24–43.

Amosz, Comenius (1659), *Orbis sensualium pictus: hoc est, omnium fundamentalium in mundo rerum, & in vita actionum, pictura & nomenclatura*, London: Swall.

Anstett, Jean-Jaques, ed. ([1804/5] 1964), *Kritische Friedrich-Schlegel-Ausgabe, vol. 12: Philosophische Vorlesungen (1800–1807) I*, München/ Zürich: Ferdinand Schöningh/Thomas.

Arciszewska, Barbara (2010), *Classicism and Modernity: Architectural Thought in Eighteenth-Century Britain*, Warsaw: Wydawnictwo Neriton.

Behler, Ernst, ed. ([1796–8] 1967), *Kritische Friedrich-Schlegel-Ausgabe, vol. 18: Philosophische Lehrjahre 1796–1806 nebst philosophischen Kritiken aus den Jahren 1796–1828*, München/ Zürich: Ferdinand Schöningh/Thomas.

Bellhouse, David R. (2017), *Leases for Lives: Life Contingent Contracts and the Emergence of Actuarial Science in Eighteenth-Century Britain*, Cambridge: Cambridge University Press.

Bernet, Jacques, ed. (2000), *Journal d'un maître d'école d'Île-de-France (1771–1792): Silly-en-Multien, de l'Ancien Régime à la Révolution*, Villeneuve d'Ascq: Presses universitaires du Septentrion.

Bernoulli, Johann (1697), "Curvatura radii in diaphanis non uniformibus", *Acta Eruditorum*, May 1697: 206-2011.

Bertoloni Meli, Domenico (1993), *Equivalence and Priority: Newton versus Leibniz*, New York: Oxford University Press.

Bertoloni Meli, Domenico (2016), "Newton and the Leibniz–Clarke Correspondence," in Rob Iliffe and George E. Smith eds., *The Cambridge Companion to Newton*, 2nd edn, 586–96, Cambridge: Cambridge University Press.

Betts, Jonathan (2004), "Finding and Keeping Time: From Stars to Pendulum," in Derek Roberts, ed., *Precision Pendulum Clocks*, 9–13, Atglen, PA: Schiffer Publishing.

Biggs, Nick (2011), "Applicable Mathematics in the 18th Century: An Example from the Textile Trade," *Mathematics Today*, 47 (2): 96–100.

Blair, Ann, Paul Duguid, Anja-Silvia Goeing, and Grafton Anthony, eds., (2021), *Information: A Historical Companion*, Princeton, NJ: Princeton University Press.

Bloor, David (1991), *Knowledge and Social Imagery*, 2nd edn, Chicago, IL: University of Chicago Press.
Bos, Henk (1974), "Differentials, Higher-Order Differentials and the Development of Leibnizian Calculus," *Archive for History of Exact Sciences*, 14 (1): 1–90.
Bosse, Heinrich (2012), *Bildungsrevolution 1770–1830*, Heidelberg: Universitätsverlag Winter.
Bottazzini, Umberto, and Amy Dahan Dalmedico, eds., (2001), *Changing Images in Mathematics: From the French Revolution to the New Millennium*, London: Routledge.
Brewer, John (1989), *The Sinews of Power: War, Money and the English State 1688–1783*, London: Routledge.
Brian, Eric (1994), *La mesure de l'État: Administrateurs et géomètres au XVIIIe siècle*, Paris: Albin Michel.
Buchwald, Jed Z., and I. Bernard Cohen, eds., (2001), *Isaac Newton's Natural Philosophy*, Cambridge, MA: MIT Press.
Buchwald, Jed Z., and Mordechai Feingold (2013), *Newton and the Origin of Civilization*, Princeton, NJ: Princeton University Press.
Buchwald, Jed Z., and Robert Fox, eds., (2013), *The Oxford Handbook of the History of Science*, Oxford: Oxford University Press.
Bullynck, Maarten (2008), "The Transmission of Numeracy: Integrating Reckoning in Protestant North-German Education (1770–1810)," *Paedagogica Historica*, 44 (5): 463–85.
Bullynck, Maarten (2010a), "A History of Factor Tables with Notes on the Birth of Number Theory 1657–1817," *Revue d'histoire des mathématiques*, 16 (2): 133–216.
Bullynck, Maarten (2010b), "Johann Lambert's Scientific Tool Kit: Exemplified by the Measurement of Humidity," *Science in Context*, 23 (1): 65–89.
Burke, Peter, and Halil Inalcik, eds., (1999), *History of Humanity, Scientific and Cultural Development, vol. 5: From the Sixteenth to the Eighteenth Century*, New York: UNESCO.
Calero, Julían Simón (2008), *The Genesis of Fluid Mechanics, 1640–1780*, Heidelberg: Springer.
Calinger, Ronald S. (2016), *Leonhard Euler. Mathematical Genius in the Enlightenment*, Princeton, NJ: Princeton University Press.
Cantor, Geoffrey N., and Michael J.S. Hodge, eds., (1981), *Conceptions of Ether: Studies in the History of Ether Theories*, Cambridge: Cambridge University Press.
Carson, Cathryn (1995), "Who Wants a Postmodern Physics?," *Science in Context*, 8 (4): 635–55.
Cavazza, Marta (2019), "Women Readers of Newton in Eighteenth-Century Italy," in Helmut Pulte and Scott Mandelbrote eds., *The Reception of Isaac Newton in Europe*, vol. 3, 841–59, London: Bloomsbury.
Chaouli, Michel (2002), *The Laboratory of Poetry: Chemistry and Poetics in the Work of Friedrich Schlegel*, Baltimore, MD: Johns Hopkins University Press.
Chrimes, Michael M., ed. (1997), *The Civil Engineering of Canals and Railways before 1850*, Aldershot: Ashgate.
Cohen, I. Bernard (1971), *Introduction to Newton's* "Principia," Cambridge: Cambridge University Press.
Cohen, I. Bernard (1999), "A Guide to Newton's *Principia*," in Isaac Newton [1726] 1999. *The* Principia: *Mathematical Principles of Natural Philosophy*, trans. I. Bernard Cohen and Anne Whitman, 1–370, Berkeley, CA: University of California Press.

Conder, Tony (2017), *Canals in Britain*, London: Shire Publications.
Condorcet, Jean-Antoine-Nicolas de Caritat ([1794] 1795), *Outlines of an Historical View of the Progress of the Human Mind*, London: J. Johnson.
Corry, Leo (1989), “Linearity and Reflexivity in the Growth of Mathematical Knowledge,” *Science in Context*, 3 (2): 409–40.
Croarken, Mary (2002), “Providing Longitude for All: The Eighteenth Century Computers of the Nautical Almanac,” *Journal for Maritime Research*, 4 (1): 106–26.
d’Alembert, Jean le Rond (1743), *Traité de dynamique*, Paris: David.
d’Alembert, Jean le Rond (1752), *Essai d’une nouvelle théorie de la résistance des fluides*, Paris: David.
d’Alembert, Jean-Baptiste le Rond ([1751] 2009), “Preliminary Discourse,” in *The Encyclopedia of Diderot & d’Alembert Collaborative Translation Project*, trans. Richard N. Schwab and Walter E. Rex, Ann Arbor, MI: Michigan Publishing, University of Michigan Library. Available online: http://hdl.handle.net/2027/spo.did2222.0001.083 (accessed April 5, 2021).
Darnton, Robert (1999), “Philosophers Trim the Tree of Knowledge: The Epistemological Strategy of the *Encyclopédie*,” in *The Great Cat Massacre: And Other Episodes in French Cultural History*, 191–213, New York: Basic Books.
Darrigol, Olivier (2005), *Worlds of Flow: A History of Hydrodynamics from the Bernoullis to Prandtl*, Oxford: Oxford University Press.
Darrigol, Olivier (2007), “The Acoustic Origins of Harmonic Analysis,” *Archive for History of Exact Sciences*, 61 (4): 343–424.
Daston, Lorraine (1988), *Classical Probability in the Enlightenment*, Princeton, NJ: Princeton University Press.
Daston, Lorraine (1994), “Enlightenment Calculations,” *Critical Inquiry*, 21 (1): 182–202.
Daston, Lorraine (1998), “Fear and Loathing of the Imagination in Science,” *Daedalus*, 127 (1): 73–95.
Davids, Karel (2008), *The Rise and Decline of Dutch Technological Leadership*, 2 vols, Leiden: Brill.
de Chadarevian, Soraya, and Nick Hopwood, eds., (2004), *Models: The Third Dimension of Science*, Stanford, CA: Stanford University Press.
de Ridder-Symoens, Hilde, ed. (1996), *A History of the University in Europe, vol. II: Universities in Early Modern Europe (1500–1800)*, Cambridge: Cambridge University Press.
De Risi, Vincenzo (2016), “The Development of Euclidean Axiomatics: The Systems of Principles and the Foundations of Mathematics in Editions of the Elements in the Early Modern Age,” *Archive for History of Exact Sciences*, 70 (6): 591–676.
Décultot, Elisabeth, ed. (2003), *Lire, copier, écrire: Les bibliothèques manuscrites et leurs usages au XVIIIe siècle*, Paris: CNRS éditions.
Deniss, John (2009), “Learning Arithmetic: Textbooks and their Users in England 1500–1900,” in Eleanor Robson and Jacqueline Stedall eds., *Oxford Handbook for the History of Mathematics*, 448–67, Oxford: Oxford University Press.
Desaguliers, J. T. (1745), *A Course of Experimental Philosophy*, vol, 1. London: W. Innys.
Desrosières. Alain (1998), *The Politics of Large Numbers*, trans. Camille Naish, Cambridge, MA: Harvard University Press.
Ducheyne, Steffen (2012), *The Main Business of Natural Philosophy: Isaac Newton’s Natural Philosophical Methodology*, Berlin: Springer.

Ducheyne, Steffen, and Jip van Besouw (2021), "Readers of the First Edition of Newton's *Principia* on the Relation between Gravity, Matter, and Divine and Natural Causation: British Public Debates, 1687–1713," *Centaurus*, 63 (2): 381–95.

Dugas, René ([1955] 1988), *A History of Mechanics*, New York: Dover.

Dunn, Richard, and Rebekah Higgitt, eds., (2015), *Navigational Enterprises in Europe and its Empires, 1730–1850*, London: Palgrave Macmillan.

Dyck, Martin (1960), *Novalis and Mathematics: A Study of Friedrich von Hardenberg's Fragments on Mathematics and its Relation to Magic, Music, Religion, Philosophy, Language and Literature*, Chapel Hill, NC: University of North Carolina Press.

Dyck, Martin (1980), "Mathematics and Literature in the German Enlightenment: Abraham Gotthelf Kästner (1719–1800)," in *Transactions of the Fifth International Congress on the Enlightenment 1980*, vol. 192, 508–12, Oxford: Voltaire Foundation.

Eckert, Michael (2002), "Euler and the Fountains of Sanssouci," *Archive for History of Exact Sciences*, 56 (6): 451–68.

Eckert, Michael (2005), *The Dawn of Fluid Dynamics: A Discipline between Science and Technology*, Hoboken, NJ: Wiley.

Edney, Matthew H., and Mary S. Pedley, eds., (2019), *The History of Cartography, vol. 4: Cartography in the European Enlightenment*, Chicago, IL: University of Chicago Press.

Eichner, Hans, ed. ([1798] 1967), *Kritische Friedrich-Schlegel-Ausgabe, vol. 2: Charakteristiken und Kritiken I (1796–1801)*, München/Zürich: Ferdinand Schöningh/Thomas.

Eichner, Hans, ed. ([n.d.] 1981), *Kritische Friedrich-Schlegel-Ausgabe, vol. 16: Fragmente zur Poesie und Literatur I*, 83–190, Paderborn/Zürich: Ferdinand Schöningh/Thomas.

Ellerton, Nerida F., and M. A. (Ken) Clements (2012), *Rewriting the History of School Mathematics in North America 1607–1861*, New York: Springer.

Engelsman, Steven B. (1984), *Families of Curves and the Origins of Partial Differentiation*, Amsterdam: North-Holland.

Euler Archive. A digital library dedicated to the work and life of Leonhard Euler. https://scholarlycommons.pacific.edu/euler-works/

Euler, Leonhard (1736), *Mechanica sive motus scientia analytice exposita*, 2 vols. St. Petersburg: Academiae Scientiarum.

Euler, Leonhard (1744), *Methodus inveniendi lineas curvas maximi minimive proprietate gaudentes sive solutio problematis isoperimetrici latissimo sens accepti*, Lausanne: Bousquet & Socios.

Euler, Leonhard (1745), "De la force de percussion et de sa véritable mesure," *Histoire de l'Académie Royale des Sciences et Belles-lettres de Berlin*, 1746 (1): 28–32.

Euler, Leonhard (1748), *Introductio in Analysin infinitorum*, vol. 1, Lausanne: M.-M. Bousquet.

Euler, Leonhard (1750), "Découverte d'un nouveau principe de mécanique," *Histoire de l'Académie Royale des Sciences et Belles-lettres de Berlin*, 1752 (6): 185–217.

Euler, Leonhard ([1750] 2015), *Reflections on Space and Time*, trans. Michael P. Sacloco and Peter Wake. Available online: http://eulerarchive.maa.org/docs/translations/E149tr.pdf (accessed May 14, 2021).

Fara, Patricia (2003), *Newton: The Making of Genius*, London: Picador.

Favre, Adrien (1931), *Les origines du systeme métrique*, Paris: Fayard.

Feingold, Mordechai (2004), *The Newtonian Moment: Science and the Making of Modern Culture*, Oxford: Oxford University Press.
Ferreiro, Larrie D. (2007), *Ships and Science: The Birth of Naval Architecture in the Scientific Revolution, 1600–1800*, Cambridge, MA: MIT Press.
Ferreiro, Larrie D. (2010–11), "Pierre Bouguer and the Solid of Least Resistance," *Revue d'histoire des sciences*, 63 (1): 93–119.
Frangsmyr, Tore, John L. Heilbron, and Robin E. Rider, eds., (1990), *The Quantifying Spirit in the Eighteenth Century*, Berkeley, CA: University of California Press.
Friedman, Michael (2019), "Newton and Kant on Absolute Space: From Theology to Transcendental Philosophy," in Helmut Pulte and Scott Mandelbrote eds., *The Reception of Isaac Newton in Europe*, vol. 3, 891–907, London: Bloomsbury.
Gabbey, Alan (1980), "Force and Inertia in the Seventeenth Century: Descartes and Newton," in Stephen Gaukroger, ed., *Descartes: Philosophy, Mathematics and Physics*, 230–320, Brighton: Harvester Press.
Galilei, Galileo ([1623] 1960), "The Assayer," trans. Stillman Drake, in Galileo Galilei, Horatio Grassi, Mario Guiducci, and Johann Kepler, *The Controversy on the Comets of 1618*, trans. Stillman Drake and C. D. O'Malley, 151–336, Philadelphia, PA: University of Pennsylvania Press.
Galilei, Galileo, Horatio Grassi, Mario Guiducci, and Johann Kepler (1960), *The Controversy on the Comets of 1618*, trans. Stillman Drake and C. D. O'Malley, Philadelphia, PA: University of Pennsylvania Press.
Gataker, Thomas (1619), *Of the Nature and Use of Lots: A Treatise Historicall and Theologicall*, London: Griffin.
Gaukroger, Stephen, ed. (1980), *Descartes: Philosophy, Mathematics and Physics*, Brighton: Harvester Press.
Gerbino, Anthony, and Stephen Johnston (2009), *Compass and Rule: Architecture as Mathematical Practice in England, 1500–1750*, New Haven, CT: Yale University Press.
Gilain, Christian, and Guilbaud Alexandre, eds. (2015), *Sciences mathematiques 1750–1850: Continuites et ruptures*, Paris: CNRS éditions.
Goethe, Johann Wolfgang ([n.d.] 1993), *Sämtliche Werke, Briefe, Tagebücher und Gespräche, vol. I.13: Sprüche in Prosa. Sämtliche Maximen und Reflexionen*, ed. Harald Fricke, Frankfurt am Main: Deutscher Klassiker Verlag.
Goethe, Johann Wolfgang ([1810] 1840), *Theory of Colours*, trans. Charles Lock Eastlake, London: John Murray.
Goethe, Johann Wolfgang ([1826] 1989), "Über Mathematik und deren Mißbrauch so wie das periodische Vorwalten einzelner wissenschaftlicher Zweige." In *Sämtliche Werke, Briefe, Tagebücher und Gespräche, vol. I.25: Schriften zur Allgemeinen Naturlehre, Geologie und Mineralogie*, ed. Wolf von Engelhardt and Manfred Wenzel, 65–76, Frankfurt am Main: Deutscher Klassiker Verlag.
Goethe, Johann Wolfgang ([1829] 1995), *Wilhelm Meister's Journeyman Years or the Renunciants*, trans. Krishna Winston, in *Goethe's Collected Works*, vol. 10, ed. Jane K. Brown, 93–436, Princeton, NJ: Princeton University Press.
Goldenbaum, Ursula (2008), "Indivisibilia Vera—How Leibniz Came to Love Mathematics," in Ursula Goldenbaum and Douglas Jesseph eds., *Infinitesimal Differences: Controversies between Leibniz and His Contemporaries*, 53–76, Berlin: de Gruyter.
Goldstine, Herman H. (1980), *A History of the Calculus of Variations: From the 17th through the 19th Century*, New York: Springer.

Grabiner, Judith V. (1998), "'Some Disputes of Consequence': Maclaurin among the Molasses Barrels," *Social Studies of Science*, 28 (1): 139–68.
Grattan-Guinness, Ivor (1990), *Convolutions in French Mathematics, 1800–1840: From the Calculus and Mechanics to Mathematical Analysis and Mathematical Physics*, 3 vols, Basel: Birkhäuser.
Gray, John (1731), *Treatise on Gunnery*, London: Printed for William Innys.
Guerrini, Anita (1986), "The Tory Newtonians: Gregory, Pitcairne, and Their Circle," *Journal of British Studies*, 25 (3): 288–311.
Guicciardini, Niccolò (1989), *The Development of Newtonian Calculus in Britain 1700–1800*, Cambridge: Cambridge University Press.
Guicciardini, Niccolò (1999), *Reading the* Principia: *The Debate on Newton's Mathematical Methods for Natural Philosophy from 1687 to 1736*, Cambridge: Cambridge University Press.
Guicciardini, Niccolò (2009), *Isaac Newton on Mathematical Certainty and Method*, Cambridge, MA: MIT Press.
Guicciardini, Niccolò (2015), "Editing Newton in Geneva and Rome: The Annotated Edition of the *Principia* by Calandrini, Le Seur and Jacquier," *Annals of Science*, 72 (3): 337–80.
Guicciardini, Niccolò (2018a), *Isaac Newton and Natural Philosophy*, London: Reaktion Books.
Guicciardini, Niccolò (2018b), "The Newton-Leibniz Calculus Controversy, 1708–1730," in Eric Schliesser and Chris Smeenk eds., *The Oxford Handbook of Newton*, Oxford: Oxford Academic. Available online: https://doi.org/10.1093/oxfordhb/9780199930418.013.9 (accessed May 17, 2023).
Guyer, Paul (2000), *Kant on Freedom, Law, and Happiness*, Cambridge: Cambridge University Press.
Hacking, Ian (1990), *The Taming of Chance*, Cambridge: Cambridge University Press.
Hadfield, C. (1997), "Rivers and Canals," in Michael M. Chrimes, ed., *The Civil Engineering of Canals and Railways before 1850*, 49–80, Aldershot: Ashgate.
Hahn, Roger (1971), *The Anatomy of a Scientific Institution: The Paris Academy of Sciences, 1666–1803*, Berkeley, CA: University of California Press.
Hahn, Roger (1975), "Scientific Research as an Occupation in Eighteenth-Century Paris," *Minerva*, 13 (4): 501–13.
Hald, A. (1990), *A History of Probability and Statistics and Their Applications before 1750*, New York: Wiley.
Hall, A. Rupert (1980), *Philosophers at War: The Quarrel between Newton and Leibniz*, Cambridge: Cambridge University Press.
Hall, A. Rupert, and Norman Smith, eds., (1976), *History of Technology*, London: Mansell.
Hall, Bryan (2011), *The Arguments of Kant's Critique of Pure Reason*, with the assistance of Mark Black and Matt Sheffield, Lanham, MD: Lexington Books.
Hankins, Thomas L. (1967), "The Reception of Newton's Second Law of Motion in the Eighteenth Century," *Archives Internationales d'Histoire des Sciences*, 20: 43–65.
Hankins, Thomas L. (1997), *Science and the Enlightenment*, Cambridge: Cambridge University Press.
Hans, Nicholas (1951), *New Trends in Education in the Eighteenth Century*, London: Routledge & Kegan Paul.
Harris, Steven J. (1999), "The Study of Nature and the Universe," in Peter Burke and Halil Inalcik eds., *History of Humanity, Scientific and Cultural Development, vol. 5: From the Sixteenth to the Eighteenth Century*, 81–95, New York: UNESCO.

Hedges, Stephen A. (1978), "Dice Music in the Eighteenth Century," *Music & Letters*, 59 (2): 180–7.
Heilbron, John Lewis (1979), *Electricity in the 17th and 18th Centuries: A Study of Early Modern Physics*, Berkeley, CA: University of California Press.
Heyman, Jacques (1976), "Couplet's Engineering Memoires 1726–1733," in A. Rupert Hall and Norman Smith eds., *History of Technology*, 21–44, London: Mansell.
Heyman, Jacques (1982), *The Masonry Arch*, Chichester: Horwood.
Home, Roderick W. (2019), "The Persistence of Mechanical Imagery: Electricity and Magnetism in a Newtonian World," in Helmut Pulte and Scott Mandelbrote eds., *The Reception of Isaac Newton in Europe*, vol. 1, 479–500, London: Bloomsbury.
Howse, Derek (1989), *Nevil Maskelyne: The Seaman's Astronomer*, Cambridge: Cambridge University Press.
Hutton, Sarah (2019), "Émilie du Châtelet," in Helmut Pulte and Scott Mandelbrote eds., *The Reception of Isaac Newton in Europe*, vol. 3, 769–75, London: Bloomsbury.
Iliffe, Rob (2017), *Priest of Nature: The Religious Worlds of Isaac Newton*, Oxford: Oxford University Press.
Iliffe, Rob, and George E. Smith, eds. (2016), *The Cambridge Companion to Newton*, 2nd edn, Cambridge: Cambridge University Press.
Israel, Jonathan I. (2006), *Enlightenment Contested: Philosophy, Modernity, and the Emancipation of Man 1670–1752*, Oxford: Oxford University Press.
Jacob, Marie (2005), "Interdire la quadrature du cercle à l'Académie: une décision autoritaire des Lumières?," *Revue d'histoire des mathématiques*, 11 (1): 89–139.
Jahnke, Hans Niels, ed. (2003), *A History of Analysis*, History of Mathematics, #24, Providence, RI/London: American Mathematical Society/London Mathematical Society.
Jolles, Frank, and Edith Höltenschmidt, eds. (2006), *Kritische Ausgabe der Vorlesungen, vol. 3: Vorlesungen über Encyclopädie*, Paderborn: Schöningh.
Kant, Immanuel ([1764] 1992), "Inquiry concerning the distinctness of the principles of natural theology and morality, being an answer to the question proposed for consideration by the Berlin Royal Academy of Sciences for the year 1763," in *Theoretical Philosophy, 1755–1770*, ed. and trans. David Walford and Ralf Meerbote, 243–75, Cambridge: Cambridge University Press.
Kant, Immanuel ([1781/1787] 1998), "Critique of Pure Reason," in *Critique of Pure Reason*, ed. and trans. Paul Guyer and Allen W. Wood, 81–704, Cambridge: Cambridge University Press.
Kant, Immanuel ([1790] 2000), *Critique of the Power of Judgement*, trans. Paul Guyer and Eric Matthews, ed. Paul Guyer, Cambridge: Cambridge University Press.
Kant, Immanuel ([1798] 2007), "Anthropology from a Pragmatic Point of View," trans. Robert B. Louden, in *Anthropology, History, and Education*, ed. Robert B. Louden and Günter Zöller, 227–429, Cambridge: Cambridge University Press.
Karp, Alexander, and Gerd Schubring (2014), *Handbook on the History of Mathematics Education*, New York: Springer.
Kästner, Abraham Gotthelf (1747), "Das Lob der Sternkunst." *Hamburgisches Magazin, oder gesammlete Schriften zum Unterricht und Vergnügen aus der Naturforschung und den angenehmen Wissenschaften überhaupt*, 1 (2): 206–22.
Kästner, Abraham Gotthelf (1786), *Fortsetzung der Rechenkunst in Anwendung auf mancherley Geschäffte*, Göttingen: Vandenhoek.
Katz, Victor J. (1995), "Ideas of Calculus in Islam and India," *Mathematics Magazine*, 68 (3): 163–74.

Katz, Victor J. (2009), *A History of Mathematics: An Introduction*, 3rd edn, Boston, MA: Pearson.

Kitcher, Philip (1992), "Kant and the Foundations of Mathematics," in Carl J. Posy, ed., *Kant's Philosophy of Mathematics: Modern Essays*, 109–31, Dordrecht: Kluwer.

Klehn, Susanne (1999), "Alphabetisierung in der Stadt Braunschweig um 1800: Die Magnigemeinde als Stichprobe," in Hans Erich Bödeker and Ernst Hinrichs eds., *Alphabetisierung und Literalisierung in Deutschland in der Frühen Neuzeit.* Wolfenbütteler Studien zur Aufklärung, #26, 141–62, Tübingen: De Gruyter.

Kline, Morris (1960), *Mathematics and the Physical World*, London: John Murray.

Kline, Morris (1972), *Mathematical Thought from Ancient to Modern Times*, New York: Oxford University Press.

Koldewey, Friedrich (1898), *Braunschweigische Schulordnungen von den ältesten Zeiten bis zum Jahre 1828*, Berlin: Hofmann.

Kuhn, Thomas S. 1976. "Mathematical vs. Experimental Traditions in the Development of Physical Science." *Journal of Interdisciplinary History*, 7, (1): 1–31.

Kuhn, Thomas S. (1977), *The Essential Tension: Selected Studies in Scientific Tradition and Change*, Chicago, IL: University of Chicago Press.

Kuhn, Thomas S. (1996), *The Structure of Scientific Revolutions*, 3rd edn, Chicago, IL: University of Chicago University Press.

Lagrange, Joseph Louis (1788), *Méchanique Analitique*, Paris: Desaint.

Lagrange, Joseph Louis (1867–92), *Oeuvres de Lagrange*, ed. Joseph Alfred Serret and Gaston Darboux, 14 vols, Paris: Gauthier-Villars.

Lalande, Jérôme de (1774), *Abrégé d'Astronomie*, Amsterdam: Barthelemie Vlam.

Lalande, Jerôme de (1803), *Supplémens au Dictionnaire des athées anciens et modernes de Sylvain Maréchal*, Paris: Lalande.

Lambert, Johann Heinrich (1765), "Discours de reception," *Histoire de l'Académie royale des sciences et belles-lettres*, année 1765, 506–14.

Landes, David S. (1983), *Revolution in Time: Clocks and the Making of the Modern World*, Cambridge, MA: Harvard University Press.

Laplace, Pierre-Simon (1796) *Exposition du système du monde*, 2 vols., Paris: Imprimerie du Cercle-Social.

Laplace, Pierre-Simon (1799-1805). *Traité de mécanique céleste*, 4 vols., Paris: Duprat (vols. 1–3), Courcier (vol.4).

Laplace, Pierre-Simon (1814), *Essai philosophique sur les probabilités*, Paris: Courcier.

Laursen, John Christian (1999), "The Berlin Academy," in Richard H. Popkin, ed., *The Columbia History of Western Philosophy*, 490–4, New York: Columbia University Press.

Leibniz, Gottfried Wilhelm (1684), "Nova methodus pro maximis et minimis.", *Acta eruditorum*, 1684, 467–73.

Leibniz, Gottfried Wilhelm ([1675] 1989), "Letter to Henry Oldenburg," in *Philosophical Papers and Letters*, 2nd edn, ed. and transl. Leroy E. Loemker, 165–6, Dordrecht: Kluwer.

Leibniz, Gottfried Wilhelm ([1677] 1989), "Dialogue," in *Philosophical Papers and Letters*, 2nd edn, ed. and transl. Leroy E. Loemker, 182–5, Dordrecht: Kluwer.

Levi, Enzo (1995), *The Science of Water: The Foundation of Modern Hydraulics*, New York: ASCE Press.

Locke, John (1706a), "Of the Conduct of the Understanding," in *Posthumous Works of Mr. John Locke*, 1–137, London: Churchill.

Loemker, Leroy E. (1989), "Introduction: Leibniz as Philosopher," in Gottfried Wilhelm Leibniz, *Philosophical Papers and Letters*, 2nd edn, ed. and trans. Leroy E. Loemker, 1–69, Dordrecht: Kluwer.

Love, John (1688), *Geodæsia, or, The art of surveying and measuring of land made easie by plain and practical rules, how to survey, protract, cast up, reduce or divide any piece of land whatsoever: with new tables for the ease of the surveyor in reducing the measures of land: moreover, a more facile and sure way of surveying by the chain, than has hitherto been taught: as also, how to lay-out new lands in America, or elsewhere: and how to make a perfect map of a river's mouth or harbour: with several other things never yet publish'd in our language*. London: Taylor.

Mach, Ernst ([1933] 1982), *Die Mechanik, historisch-kritisch dargestellt*, Darmstadt: Wissenschaftliche Buchgestellschaft.

Malet, Antoni (2020), "Newton's Mathematics," in Eric Schliesser and Chris Smeenk eds., *The Oxford Handbook of Newton*, Oxford: Oxford Academic. Available online: https://doi.org/10.1093/oxfordhb/9780199930418.013.24 (accessed May 17, 2023).

Martin, Benjamin (1782), *The Mariner's Mirror*, London: Robert Sayer.

McGee, David (1999), "From Craftsmanship to Draftsmanship: Naval Architecture and the Three Traditions of Early Modern Design," *Technology and Culture*, 40 (2): 209–36.

McGrayne, Sharon Bertsch (2011), *The Theory that Would Not Die: How Bayes' Rule Cracked the Enigma Code, Hunted Down Russian Submarines & Emerged Triumphant from Two Centuries of Controversy*, New Haven, CT: Yale University Press.

McMurran, Shawnee, and V. Frederick Rickey (2008), "The Impact of Ballistics on Mathematics," *Proceedings of the 16th ARL/USMA Technical Symposium*, 1–25.

Mendelssohn, Moses ([1764] 1997), "On Evidence in Metaphysical Sciences," in *Philosophical Writings*, ed. and trans. Daniel O. Dahlstrom, 253–306, Cambridge: Cambridge University Press.

Morel, Thomas (2014), "Arithmetical Periodicals in Late Eighteenth-Century Germany: 'Mathematics for the Use and the Pleasure of Civic Life,'" *Archives Internationales d'Histoire des Sciences*, 64: 172–3, 397–427.

Morel, Thomas, and Maarten Bullynck (2015), "Une revolution peut en cacher d'autres. Le paysage morcele des mathematiques dans l'espace germanophone et ses reconfigurations (1750–850)," in Christian Gilain and Guilbaud Alexandre eds., *Sciences mathematiques 1750–1850: Continuites et ruptures*, 181–206, Paris: CNRS éditions.

Morton, Alan Q., and Jane A. Wess (1993), *Public and Private Science: The King George III Collection*, Oxford: Oxford University Press.

Neve, Richard (1708), *Mathematicks Made Plain, in the Solution of a Variety of Useful Propositions in Arithmetic, & c.*, London: Conyers, Sprint & Ballard.

Neves, Marcelo Almeida Santos, Vadim L. Belenky, Jean Otto de Kat, Kostas Spyrou, and Naoya Umeda, eds. (2011), *Contemporary Ideas on Ship Stability and Capsizing in Waves*, Dordrecht: Springer.

Newton, Isaac (1704), *Opticks: or, a Treatise of the Reflexions, Refractions, Inflexions and Colours of Light*, London: Printed for Sam. Smith and Benj. Walford.

Newton, Isaac ([1726] 1999), *The Principia: Mathematical Principles of Natural Philosophy*, trans. I. Bernard Cohen and Anne Whitman, 3rd edn, Berkeley, CA: University of California Press.

Newton, Isaac (1959–77), *The Correspondence*, ed. A. Rupert Hall, Laura Tilling, Joseph F. Scott, and Herbert Westren Turnbull, 7 vols, Cambridge: Cambridge University Press.

Newton, Isaac (1967–81), *The Mathematical Papers of Isaac Newton*, ed. D. T. Whiteside with the assistance of M. A. Hoskin, 8 vols, Cambridge: Cambridge University Press.

Novalis ([1798–99] 2007), *Notes for a Romantic Encyclopedia. Das Allgemeine Brouillon*, ed. and trans. David W. Wood, Albany, NY: State University of New York Press.

Novalis ([1802] 1842), *Henry of Ofterdingen: A Romance*, Cambridge: John Owen.

Nowacki, Horst (2008), "Leonhard Euler and the Theory of Ships," *Journal Ship Research*, 52 (4): 274–90.

Nowacki, Horst, and Larrie D. Ferreiro (2011), "Historical Roots of the Theory of Hydrostatic Stability of Ships," in Marcelo Almeida Santos Neves, Vadim L. Belenky, Jean Otto de Kat, Kostas Spyrou, and Naoya Umeda eds., *Contemporary Ideas on Ship Stability and Capsizing in Waves*, 141–80, Dordrecht: Springer.

Orozco-Echeverri, Sergio H., and Sebastián Molina-Betancur (2019), "José Celestino Mutis' Appropriation of Newton's Experimental Physics in New Granada (1761–1808)," *History of Science*, 57 (3): 291–323.

Peiffer, Jeanne, and Jean-Pierre Vittu, eds. (2013), "Les journaux savants dans l'Europe des XVIIe et XVIIIe siecles. Formes de la communication et agents de la construction des savoirs," *Archives internationales d'histoire des sciences*, 63 (June/December): 170–1.

Pesic, Peter (2013), "Euler's Musical Mathematics," *Mathematical Intelligencer*, 35 (2): 35–43.

Pestalozzi, Johann Heinrich (1801), *Wie Getrud ihre Kinder lehrt, ein Versuch den Müttern Anleitung zu geben, ihre Kinder selbst zu unterrichten*, Bern: Gessler.

Popkin, Richard H., ed. (1999), *The Columbia History of Western Philosophy*, New York: Columbia University Press.

Posy, Carl J., ed. (1992), *Kant's Philosophy of Mathematics: Modern Essays*, Dordrecht: Kluwer.

Pulte, Helmut (1989), *Das Prinzip der kleinsten Wirkung und die Kraftkonzeptionen der rationalen Mechanik. Eine Untersuchung zur Grundelegungsproblematik bei Leonhard Euler, Pierre Louis Moreau de Maupertuis und Joseph Louis Lagrange*, Stuttgart: Franz Steiner.

Pulte, Helmut (2005), *Axiomatik und Empirie. Eine wissenschaftstheoriegeschichtliche Untersuchung zur Mathematischen Naturphilosophie von Newton bis Neumann*, Darmstadt: Wissenschaftliche Buchgesellschaft.

Pulte, Helmut, and Scott Mandelbrote, eds. (2019), 3 vols., *The Reception of Isaac Newton in Europe*, London: Bloomsbury Academic.

Pycior, Helena (1997), *Symbols, Impossible Numbers, and Geometric Entanglements: British Algebra through the Commentaries on Newton's Universal Arithmetick*, Cambridge: Cambridge University Press.

Remer, Christain S. (1737), *Demonstrativische Anweisung zur Rechen=Kunst für diejenigen, in derselben den rechten Grundlegen wollen, und welche im gemeinen Leben unentbehrlich*, Braunschweig: Schröder.

Reynolds, Terry S. (1983), *Stronger than a Hundred Men: A History of the Vertical Water Wheel*, Baltimore, MD: Johns Hopkins University Press.

Roberts, Derek, ed. (2004), *Precision Pendulum Clocks*, Atglen, PA: Schiffer Publishing.

Robson, Eleanor, and Jacqueline Stedall, eds. (2009), *The Oxford Handbook of the History of Mathematics*, Oxford: Oxford University Press.

Rozier, Abbé, ed. (1809), *Nouveau cours complet d'agriculture théorique et pratique ou Dictionnaire raisonné et universel d'agriculture*, 13 vols, Paris: Deterville.
Ruddock, Ted (2017), *Masonary Bridges, Viaducts and Aquaducts*, London: Routledge.
Rüger, Alexander (1985), "Die Molekularhypothese in der Theorie der Kapillarerscheinungen (1805–1873)," *Centaurus*, 28 (3): 244–76.
Russell, John L. (1975), "Action and Reaction before Newton," *British Journal for the Philosophy of Science*, 8: 25–38.
Sakarovitch, Joël (1998), *Épures d'architecture. De la coupe des pierres à la géométrie descriptive, XVIe–XIXe siècles*, Historical Studies, Science Networks, #21, Basel: Birkhäuser.
Sandifer, C. Edward (2007), *The Early Mathematics of Leonhard Euler*, Washington, DC: Mathematical Association of America.
Sarret, Jean-Baptiste (1799), *Observations pour les instituteurs sur les élémens d'arithmétique à l'usage des écoles primaires*, Paris: Didot et Deterville.
Schaffer, S. (2004), "Fish and Chips: Models in the Age of Reason," in Soraya de Chadarevian and Nick Hopwood eds., *Models: The Third Dimension of Science*, 71–109, Stanford, CA: Stanford University Press.
Schlegel, Friedrich ([n.d.] 1981), "Fragmente zur Litteratur und Poesie," in Hans Eichner, ed., *Kritische Friedrich-Schlegel-Ausgabe, vol. 16: Fragmente zur Poesie und Literatur I*, 83–190, Paderborn/Zürich: Ferdinand Schöningh/Thomas.
Schlegel, Friedrich ([1796–8] 1967), "Philosophische Fragmente. Erste Epoche II," in Ernst Behler, ed., *Kritische Friedrich-Schlegel-Ausgabe, vol. 18: Philosophische Lehrjahre 1796–1806 nebst philosophischen Kritiken aus den Jahren 1796–1828*, 17–119, Paderborn/Zürich: Ferdinand Schöningh/Thomas.
Schlegel, Friedrich ([1798] 1967), "Athenäums-Fragmente," in Hans Eichner, ed., *Kritische Friedrich-Schlegel-Ausgabe, vol. 2: Charakteristiken und Kritiken I (1796–1801)*, 165–255, Paderborn/Zürich: Ferdinand Schöningh/Thomas.
Schlegel, August Wilhelm ([1803] 2006), *Kritische Ausgabe der Vorlesungen, vol. 3: Vorlesungen über Encyclopädie*, ed. Frank Jolles and Edith Höltenschmidt, Paderborn/Zürich: Ferdinand Schöningh/Thomas.
Schlegel, Friedrich ([1804/05] 1964), "Die Entwicklung der Philosophie in zwölf Büchern," in Jean-Jaques Anstett, ed., *Kritische Friedrich-Schlegel-Ausgabe, vol. 12: Philosophische Vorlesungen (1800–1807) I*, 107–480, Paderborn/Zürich: Ferdinand Schöningh/Thomas.
Schliesser, Eric (2013), "On Reading Newton as an Epicurean: Kant, Spinozism and the Changes to the *Principia*," *Studies in History and Philosophy of Science*, 44A: 416–28.
Schliesser, Eric, and Chris Smeenk, eds. (2018), *The Oxford Handbook of Newton*, Oxford: Oxford Academic.
Schneider, Ivo (2011), "Goethe als Vorbild für die Einstellung deutscher Bildungsbürger zur Mathematik," *Goethe-Jahrbuch*, 128: 251–62.
Schubring, Gerd (2005), *Conflicts between Generalization, Rigor, and Intuition: Number Concepts Underlying the Development of Analysis in 17–19th Century France and Germany*, Berlin: Springer.
Scott, Wilson L. (1970), *The Conflict between Atomism and Conservation Theory 1644 to 1860*, London: MacDonald.
Shank, J. B. (2008), *The Newton Wars and the Beginning of the French Enlightenment*, Chicago, IL: University of Chicago Press.
Shapiro, Alan E. (2019), "The Reception of Newton's Optical Work," in Helmut Pulte and Scott Mandelbrote eds., *The Reception of Isaac Newton in Europe*, 403–32, London: Bloomsbury.

Skempton A. W. (1996), *Civil Engineers and Engineering in Britain, 1600–1830*, Aldershot: Variorum.
Smeaton, John (1759), "An Experimental Enquiry Concerning the Natural Powers of Water and Wind to Turn Mills, and Other Machines, Depending on a Circular Motion," *Philosophical Transactions*, 51 (1759–60): 100–74.
Smeenk, Chris, and Eric Schliesser (2013), "Newton's *Principia*," in Jed Z. Buchwald and Robert Fox eds., *The Oxford Handbook of the History of Science*, 109–65, Oxford: Oxford University Press.
Smith, George E. (2001), "The Newtonian Style in Book II of the *Principia*," in Jed Z. Buchwald and I. Bernard Cohen eds., *Isaac Newton's Natural Philosophy*, 249–98, Cambridge, MA: MIT Press.
Sonar, Thomas (2018), *The History of the Priority Dispute between Newton and Leibniz: Mathematics in History and Culture*, Basel: Birkhäuser.
Stammel, Hans (1982), "Der Kraftbegriff in Leibniz' Physik," PhD diss., University of Mannheim.
Stedall, Jacqueline (2002), *A Discourse Concerning Algebra: English Algebra to 1685*, Oxford: Oxford University Press.
Stewart, L., and Weindling, P. (1995), "Philosophical Threads: Natural Philosophy and Public Experiment among the Wavers of Spitalfields," *British Journal for the History of Science*, 28 (1): 37–62.
Stigler, Stephen M. (1986), *The History of Statistics: The Measurement of Uncertainty before 1900*, Cambridge, MA: Belknap Press.
Struik, Jan Dirk (1986), *A Source Book in Mathematics, 1200–1800*, Princeton, NJ: Princeton University Press.
Swift, Jonathan (1726), *Travels into Several Remote Nations of the World. In Four Parts. By Lemuel Gulliver, First a Surgeon, and then a Captain of Several Ships*, London: Motte.
Taton, René, ed. (1964), *Enseignement et diffusion des sciences en France au dix-huitieme siècle*, Paris: Hermann.
Taylor, E. G. R. (1966), *The Mathematical Practitioners of Hanoverian England, 1714–1840*, Cambridge: Cambridge University Press for the Institute of Navigation.
Telescope, Tom (1761), *The Newtonian System of Philosophy, Adapted to the Capacities of Young Gentlemen and Ladies*, London: J. Newbery.
Terrall, Mary (2002), *The Man Who Flattened the Earth: Maupertuis and the Sciences in the Enlightenment*, Chicago, IL: University of Chicago Press.
Todhunter, Isaac ([1873] 1961), *A History of the Calculus during the Nineteenth Century*, New York: Chelsea Publishing.
Tournès, Dominique, ed. (2022), *Histoire du calcul graphique*, Paris: Cassini.
Truesdell, Clifford A. (1968), *Essays in the History of Mechanics*, Berlin: Springer.
van Besouw, Jip (2020), "Willem Jacob's Gravesande's Philosophical Trajectory: 'Between' Leibniz and Newton," *Intellectual History Review*, 30 (4): 615–40.
van Besouw, Jip, and Steffen Ducheyne (2021), "Characterisations in Britain of Isaac Newton's Approach to Physical Inquiry in the *Principia* between 1687 and 1713," *Early Science and Medicine*, 26 (4): 341–72.
Voltaire (1734), *Lettres Philosophiques*, Amsterdam: E. Lucas.
von Waltershausen, Sartorius (1856), *Gauß zum Gedächtniss*, Leipzig: Hirzel.
Wardhaugh, Benjamin (2009), "Mathematics, Music, and Experiment in Late Seventeenth-Century England," in Eleanor Robson and Jacqueline Stedall eds., *The Oxford Handbook of the History of Mathematics*, 639–61, Oxford: Oxford University Press.

Wardhaugh, Benjamin (2012), *Poor Robin's Prophecies: A Curious Almanac and the Everyday Mathematics of Georgian Britain*, Oxford: Oxford University Press.
Wess, Jane (1995), "Lecture Demonstrations and the Real World: The Case of Cart-Wheels," *British Journal for the History of Science*, 28 (1): 79–90.
Wess, Jane (2012), "Avoiding Arithmetic, or the Material Culture of Not Learning Mathematics," *BSHM Bulletin: Journal of the British Society for the History of Mathematics*, 27 (2): 82–106.
Wess, Jane (2015), "Navigation and Mathematics: A Match Made in the Heavens?," in Richard Dunn and Rebekah Higgitt eds., *Navigational Enterprises in Europe and its Empires, 1730–1850*, 201–22, London: Palgrave Macmillan.
Westfall, Richard S. (1971), *Force in Newton's Physics: The Science of Dynamics in the 17th Century*, London: MacDonald.
Westfall, Richard S. (1980), *Never at Rest: A Biography of Isaac Newton*, Cambridge: Cambridge University Press.
Wilson, Curtis A. (1985), "The Great Inequality of Jupiter and Saturn: From Kepler to Laplace," *Archive for History of Exact Sciences*, 33 (1–3): 15–290.
Wolff, Christian (1720), *Vernünfftige Gedancken von Gott, der Welt und der Seele des Menschen*, Halle: Renger.
Yeomans, David T. (1999), "Designing the Beam: From Rules of Thumb to Calculations," in William Addis, ed., *Structural and Civil Engineering Design*, 43–9, Aldershot: Ashgate.
Yoder Joella G. (1988), *Unrolling Time: Christian Huyghens and the Mathematization of Nature*, Cambridge: Cambridge University Press.
Zollmann, Johann W. (1744), *Vollständige Anleitung zur Geodäsie oder practischen Geometrie*, Halle: Renger.

INDEX

Locators in *italics* refer to illustrations